3 - 20

65 - 75

79 - 191

CELL MOVEMENT AND NEOPLASIA

CELL MOVEMENT AND NEOPLASIA

Proceedings of the annual meeting of the Cell Tissue and Organ Culture Study Group, held at the Janssen Research Foundation, Beerse, Belgium, May 1979

Editors

M. DE BRABANDER

Janssen Pharmaceutica Research Laboratories, Beerse, Belgium

M. MAREEL

Kliniek voor Radiotherapie & Kerngeneeskunde, Gent, Belgium

L. DE RIDDER

Akademisch Ziekenhuis, Gent

PERGAMON PRESS

OXFORD · NEW YORK · TORONTO · SYDNEY · PARIS · FRANKFURT

U.K.	Pergamon Press Ltd., Headington Hill Hall, Oxford OX3 0BW, England
U.S.A.	Pergamon Press Inc., Maxwell House, Fairview Park, Elmsford, New York 10523, U.S.A.
CANADA	Pergamon of Canada, Suite 104, 150 Consumers Road, Willowdale, Ontario M2J 1P9, Canada
AUSTRALIA	Pergamon Press (Aust.) Pty. Ltd., P.O. Box 544, Potts Point, N.S.W. 2011, Australia
FRANCE	Pergamon Press SARL, 24 rue des Ecoles, 75240 Paris, Cedex 05, France
FEDERAL REPUBLIC OF GERMANY	Pergamon Press GmbH, 6242 Kronberg-Taunus, Pferdstrasse 1, Federal Republic of Germany

First edition 1980

British Library Cataloguing in Publication Data

Cell Tissue and Organ Culture Study Group. *Annual Meeting, Beerse, 1979*
Cell movement and neoplasia.
1. Cancer cells - Congresses
2. Cells - Motility - Congresses
I. Title II. Brabander, Marc de
III. Mareel, M IV. Ridder, L de
616.9'94 RC269 79-42740

ISBN 0-08-025534-5

In order to make this volume available as economically and as rapidly as possible the authors' typescripts have been reproduced in their original forms. This method has its typographical limitations but it is hoped that they in no way distract the reader.

Printed in Great Britain by A. Wheaton & Co., Ltd., Exeter

Contents

Preface

Both fundamentalists and clinicians have always recognized that the major problem of cancer is the dissemination of malignant cells and the formation of metastases in the entire body. These escape the classical therapy consisting of surgery or radiotherapy. Chemotherapeutic agents, in theory at least, should be able to interfere with the process. However, the development of cancer chemotherapy has closely followed the path of molecular biology. The elucidation of the structure of DNA and the synthetic machinery of the cell was paralleled by the development of an armamentarium of drugs which aimed mainly at blocking cell proliferation. The high degree of toxicity of these drugs which limits their use and efficacy leads one to assume that the basic steps involved in cell division are quite similar in normal and malignant cells.

The pioneering work of Lewis, Abercrombie and others revived the interest in cancer cell locomotion as a most important factor in malignant invasion and metastasis. Molecular and cell-biological research in this field have expanded quickly. Today many of the molecules involved in cell locomotion have been identified, their subcellular organization has been elucidated to a great extent and some of the regulatory mechanisms are beginning to emerge. This knowledge should be of great use in devising new therapeutic approaches which aim more specifically at malignant processes such as invasion and metastasis. This is the reason why the Cell Tissue and Organ Culture study group thought it wise to devote its 1979 annual meeting to the topic of Cell Movement and Neoplasia. This volume contains the proceedings of that meeting. Scientists from diverse disciplines were gathered who contributed to the various facets of this multidisciplinary volume. The state of the art and the problems to be faced are reviewed in an excellent paper by Peter Straüli. New methods and data are introduced in such diverse fields as the subcellular distribution and assembly of mechanochemical proteins involved in locomotion; the migration of cancer cells in monolayer cultures and their invasion into non-malignant tissues in organotypical culture and in experimental animals; the effects of ionising radiation and old and new anticancer agents on cell locomotion and invasion. Finally, most relevant information is obtained from comparative investigations on cellular migration and interactions during early morphogenesis.

We sincerely hope that this volume will be as useful to the reader as the meeting was to the participants.

We wish to express our gratitude to the Janssen Research Foundation who, very generously, hosted the meeting and the following companies and institutions who participated in lightening the financial burden: Abbott, Beecham Pharmaceutical, Eli Lilly Benelux, Gibco-Biocult, I.C.I., I.C.N., Medatom, Unilever, Upjohn, Van Ermenghem, Wellcome, The Nationaal Fonds voor Wetenschappelijk Onderzoek and the Ministerie van Nationale Opvoeding en Nederlandse Kultuur.

M. De Brabander
M. Mareel
L. De Ridder

Structural Functional Correlates in Cell Movement and Invasion

Serum α_2M: a Cell Attachment Factor?

M. Brugmans and J. J. Cassiman

Division of Human Genetics, University of Leuven, Minderbroedersstraat 12, B-3000 Leuven, Belgium

ABSTRACT

The hypothesis was tested that serum components for which the cells carry a receptor might promote or affect the attachment and/or spreading of these cells onto a substratum coated with these components. For this purpose α_2Macroglobulin, for which a receptor on normal human fibroblasts was demonstrated was used (Van Leuven, 1979).
Although this molecule proved to be a good substratum for cell attachment and allowed cell spreading, no evidence could be obtained that the presence of a cell surface receptor was required for this cellular behavior.

KEYWORDS

Attachment; spreading, normal human fibroblast, tumoral cells, α_2Macroglobulin, receptor mediated.

INTRODUCTION

Abnormal growth, invasion and metastasis are properties characterizing the abnormal social behavior of malignant cells. To understand the behavior of malignant cells is to comprehend how cells interact with each other and with the environment in order to form and maintain organized tissues, and organs.
Although little is known about the mechanisms of cellular interactions, it is apparent that the cell surface mediates the cells' relation with its environment. During the transport of tumor cells in the circulating system, the cells can undergo a variety of interactions. It seems evident that the fate of these malignant cells will not only be affected by their surface composition, but also by the components present in the circulating system, the serum components.
How serum components affect the behavior of cells is being studied *in vitro*. Lately the role of serum components in the attachment and spreading has received much attention (Yamada, 1978; Grinnell, 1978; Seglen, 1978a). Attempts to isolate the active-spreading factor have converged to the glycoprotein fibronectin (Yamada, 1978; Grinnell, 1978).
These studies indicate that Fibronectin functions as an adhesive molecule for normal and tumoral cells, but how it functions and whether it is the only adhesive molecule present in serum is still unknown.

It is established on the other hand that some serum components bind to a specific surface molecule, a receptor, and that the cells react to this by redistributing their membrane components resulting in patching and/or endocytosis.
Based on these findings the hypothesis has been proposed (Rajaraman, 1978) that, when molecules, which are able to bind to a specific receptor, are adsorbed to a substratum, they will bind their receptor while endocytosis cannot occur. As a result, the cell surface receptors will redistribute, binding still more adsorbed molecules, and spreading of the cells would occur.
There is no evidence however, that receptor mediated spreading does indeed occur. Till now a surface receptor for fibronectin has indeed not been demonstrated.
To test the hypothesis that serum components for which cells have a specific surface receptor might affect the attachment or spreading of the cell onto a substratum coated with these serum components, we have used the $\alpha_2$macroglobulin-system for which a surface receptor has been demonstrated (Van Leuven, 1979).
Normal human fibroblasts (NHF) bind and ingest α_2M-protease complexes selectively. A human osteosarcoma cell line (MG 63) has markedly decreased binding and uptake of this molecule (Van Leuven, 1979).

MATERIALS AND METHODS

Cell Lines

Normal human fibroblasts (NHF) were obtained from skin biopsies of the forearm of healthy donors. All the cells were assayed between the 8th and 15th passage.
MG 63, a human osteosarcoma derived cell line, was a gift from H. Heremans (1978).
All cell lines were regularly screened for mycoplasma contamination.

Culture Conditions

Cells were routinely grown in Dulbecco's Modified Eagle's medium (DME, Gibco) with 1 g l^{-1} $NaHCO_3$ and buffered with TES (Calbiochem) and HEPES (Calbiochem) at 15 mM each, pH 7.4 (culture medium) and with 10% (v/v) heat inactivated Newborn Calf Serum (NCS, Seralab). The cultures were transferred in an early passage (6-8) to medium with 10% heat inactivated Human Serum (HuS), and further cultured in this medium. All assays were performed at least 4 days after this transfer. Pooled Human Serum was obtained from the Red Cross Blood Centre, Leuven, Belgium. The cells were kept in an atmosphere of 5% CO_2, 95% air and 100% humidity at 37° C. The cultures were divided by trypsin treatment (Difco, 1/250), one in two, twice a week.

Coating with Serum Proteins

A serum protein coat was obtained as follows : wells of a Linbro tray or round glass coverslips were incubated with the indicated concentration of the protein in serum-free medium without bicarbonate (assay medium) during 15 min at 37° C followed by 2 washes with serum-free assay medium.
Human plasma proteins :

α_2-Macroglobulin-trypsin complexes : purification of α_2M and preparation of α_2M-trypsin complexes (α_2M-T) was as described by Van Leuven (1978). α_2M-T was used at a concentration of 150 μg/ml.

Fibronectin : was obtained from Collaborative Res. Inc. and was used at a concentration of 50 μg/ml (F).

Human Albumin : was obtained from the Red Cross Blood Centre, Leuven, Belgium and used at a concentration of 2.2 mg/ml (HSA).

Rabbit anti-human-α_2M antibodies : were obtained from Dakopatts A/S, Denmark and used at a dilution 1/10 (a(α_2M)).

Attachment Rate

Cells were seeded in 75 cm^2 flasks at 1.5 x 10^6 cells per flask and cultured for 48 to 72 hrs. The medium was then replaced by culture medium, without Leucine (Gibco) and without serum and containing 10 μCi/ml ^{3}H-Leucine (Sp. Act. 1Ci/mM, The Radio Chemical Centre, Amersham) and further incubated at 37° C for 2 hrs. The labeled cells were dissociated with dispase (Grad. II, Boehringer, Mannheim) 4 μg/ml in 0.02% EDTA tris buffered saline (Calbiochem)) washed 3 times with serum-free medium, filtered through a double layer of Nytex (20 μm pore size) and plated in the wells of a Linbro tray (Flow Laboratories) on a coat of serum protein (see above) at a concentration of 8 x 10^4 cells per ml, 0.5 ml per well. The trays were floated on a waterbath at the indicated temperature. At the indicated time points, the non-attached cells were removed by vacuum aspiration. Two washes were applied. The number of cells remaining attached to the substratum were estimated by counting the radioactivity associated with the substratum (Brugmans, 1978).

The results are expressed as the percentage of cells attached (100% was determined by an incubation of 2h at 37° C. Microscopic examination revealed that over 90% of the cells had attached to the plastic at that time).

Spreading

Cells were treated as described under 'attachment rate'. The cell suspension was plated on round glass coverslips (Ø 13 mm) which were precoated with proteins. (see above).

The cells were left for 60 min at 37° C. Cells which had attached to the coverslips were either examined live or were fixed for 2h in 2.5% glutaraldehyde (Fluka, AG) in 0.1 M cacodylate buffer pH 7.4 and the coverslips were mounted on a slide. The spreading of the cells was examined under a light microscope.

SEM

Cells fixed in 2.5 % M glutaraldehyde in 0.1 M cacodylate buffer pH 7.4 were postfixed for 1h in 1% osmium tetroxide in 0.1 M phosphate buffer pH 7.4. After dehydration with a graded series of ethanols, the cells were dried by the critical point method (E-3000 Polaron) with carbon dioxide, coated with gold in a polaron sputter coater (E-5000 diode) and examined in Philips PSEM 500.

Reagents

Insulin (Sigma Co.) was used at conc. 10 μg/ml

EGF (Boehringer) was used at 50 μg/ml.

RESULTS

Attachment Rate of NHF and MG 63 Cells

Cell attachment is the first measurable step in the interaction of cells with a substratum, eventually leading to spreading, and was thus examined as a parameter of cell-substratum interaction.

Fig. 1A shows the kinetics of attachment of single NHF's to a coat of serum protein measured at 37° C over a time period of 60'. The attachment kinetics of the cells on α_2M-T and on Fibronectin were similar, reaching a plateau at 20 min when about 70% of the cells had attached. Cells plated on a coat of HSA showed a 20' period of low attachment (10%) whereafter they attached at a faster rate.

The MG 63 cells attached in the same manner to the various substrata as the NHF's (Fig. 1B) - that is : the attachment kinetics of the cells on α_2M-T and on Fibronectin were similar and occurred at a high rate. Cells plated on HSA showed a 20 to 40 min lagperiod before they started to attach.

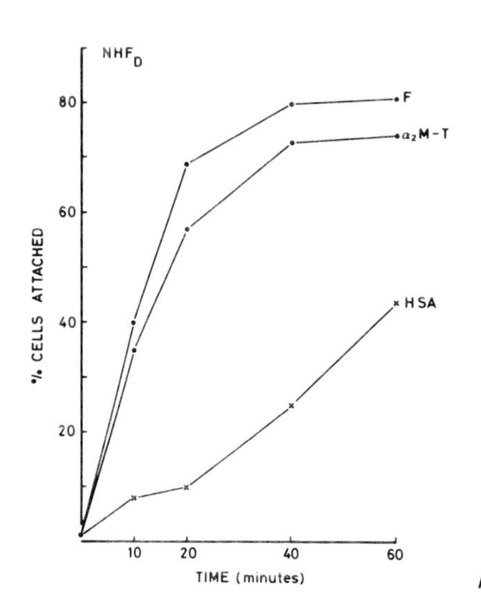

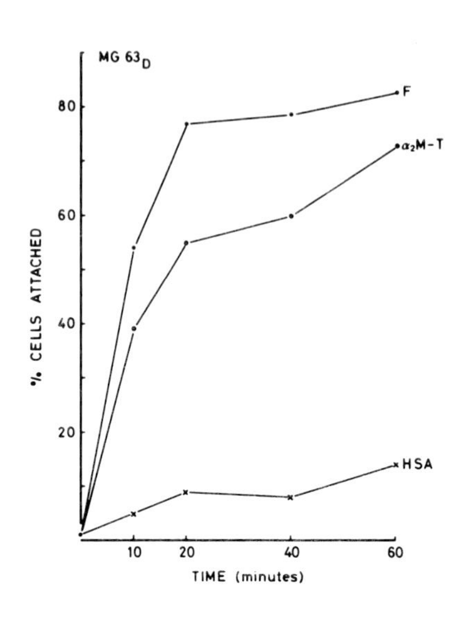

Fig. 1 A : Attachment kinetics at 37° C of dispase-dissociated NHF to a coat of various serum proteins.
Each point represents the mean of at least 4 determinations.
B : Attachment kinetics at 37° C of dispase-dissociated MG 63 to a coat of various serum proteins.
Each point represents the mean of at least 2 determinations.

The question was asked whether attachment to α_2M-T could be inhibited by the addition of free α_2M-T in the attachment medium. Competitive inhibition of attachment could not be obtained for the receptor carrying cells (NHF) nor for the 'low receptor' cells (MG 63).
Free EGF or Insulin in the medium, components known to have also a receptor on the cell surface, were also non-inhibitory. Addition of Fibronectin was also without effect on the attachment on a coat of α_2M-T.
From these experiments it can be concluded that α_2M-T is an excellent substratum for cell attachment. These results are comparable to what has been described with Fibronectin and its role in attachment of normal and transformed cells.

Temperature Dependence of the Attachment of NHF and MG 63 Cells

Plain glass or plastic, without any coating will allow fast attachment of cells. This form of attachment has been called passive, since it occurs independently of temperature and of metabolic activity. A clear understanding of temperature dependence of cell attachment has yet to be obtained, but sensitivity to temperature can be a useful parameter to distinguish different modes of attachment (Wolpert, 1969; Seglen 1978b).
Therefore it was examined whether the attachment of NHF's and MG 63 cells could be discriminated on this basis.
Figs. 2A and B illustrate the temperature dependence of attachment measured at the 60 min time point, for the two cell types. Between 4 to 37° C the extent of the attachment of NHF to a coat of α_2M-T or to Fibronectin increased gradually with the temperature. On a coat of HSA less than 10% of the cells attached within 60 min at temperatures below 22° C (Fig. 2A).
MG 63 cells demonstrated a different temperature dependence when the attachment of the cells on a coat of Fibronectin was compared to the attachment on a coat of

α_2M-T. On Fibronectin the attachment of MG 63 cells was almost temperature insensitive once 15° C was reached, whereas the attachment on α_2M-T was dependent on the temperature in a linear fashion between 4 and 37° C.

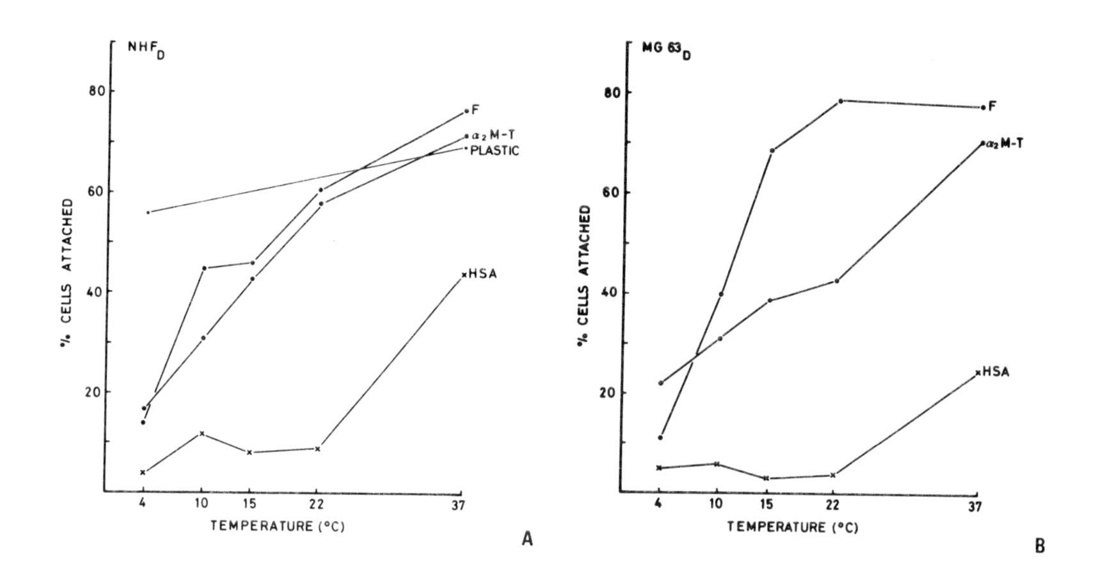

Fig. 2 A : Temperature dependence of attachment, measured 60 min after plating of NHF on a coat of serum protein.
Each point represent the mean of at least 2 determinations.
B : Temperature dependence of attachment, measured 60 min after plating of MG 63 on a coat of serum protein.
Each point represents the mean of at least 4 determinations.

Attachment on a coat of HSA was negligible at 22° C or lower (Fig. 2B).
From these results it can be concluded that the attachment of NHF and MG 63 to Fibronectin or α_2M-T does not occur in the same fashion.

Spreading of NHF and MG 63 Cells

The spreading of the cells on the various substrata was examined.
NHF plated on a coat of Fibronectin spread and obtained a fibroblastic shape whithin 60 min (Fig. 3). On α_2M-T most cells were flattening, but compared to Fibronectin the spreading was far less advanced. The Scanning Electron Micrographs (Fig. 4) show the most representative cells on each of the substrata. On HSA almost all cells were still round.
The MG 63 cells (Fig. 5) reacted quite in the same way to the different substrata. Again the cells on Fibronectin showed the greatest degree of spreading. On α_2M-T the cells were spreading; on HSA all most all cells were round. The SEMs (Fig. 6) show the most representative cells plated on a coat of α_2M-T or Fibronectin.
No competitive inhibition of spreading of NHF or MG 63 cells on α_2M-T could be obtained with α_2M-T, Fibronectin, EGF or Insulin added to the medium.
Pretreatment of the α_2M-T coat with a(α_2M) or the addition of a(α_2M) to the medium inhibited spreading.

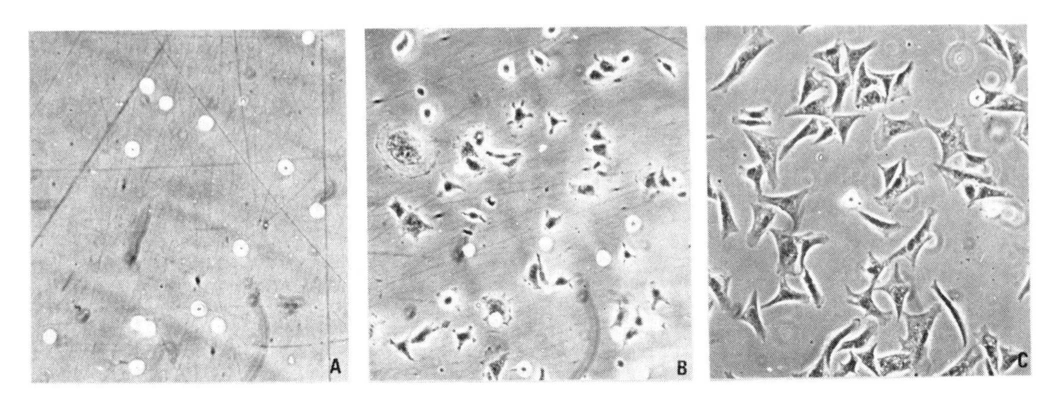

Fig. 3 : Light optic views of NHF plated on a coat of HSA (A), a_2M-T (B) or Fibronectin (C). (60 min)

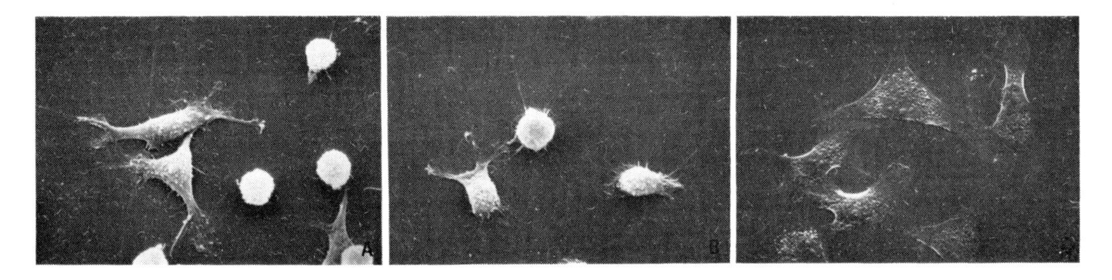

Fig. 4 : Scanning Electron Micrographs of NHF attached during 60 min on a_2M (A), a_2M-T (B) or Fibronectin (C).

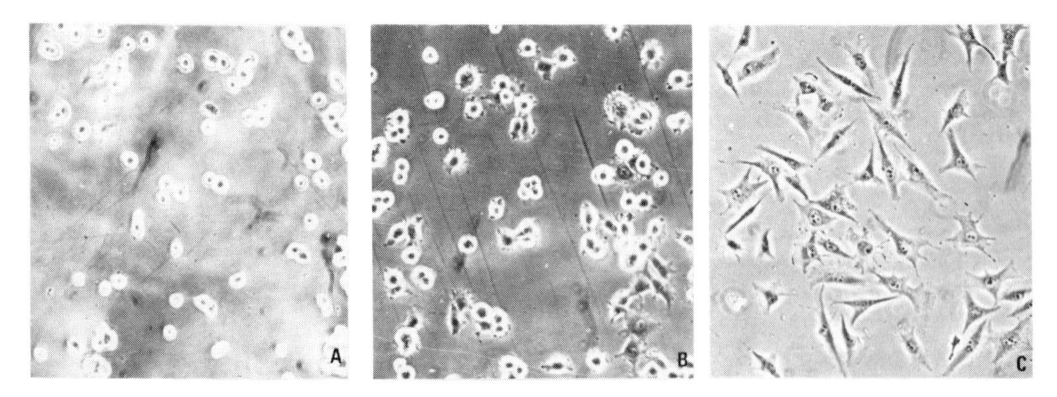

Fig. 5 : Light optic views of MG 63 plated on a coat of HSA (A), a_2M-T (B) or Fibronectin (C). (60 min)

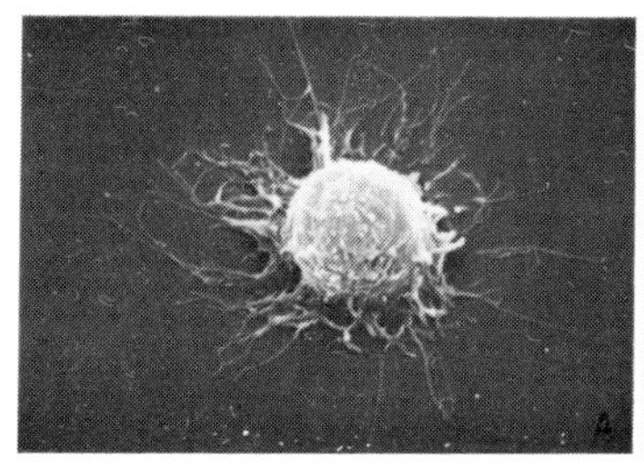
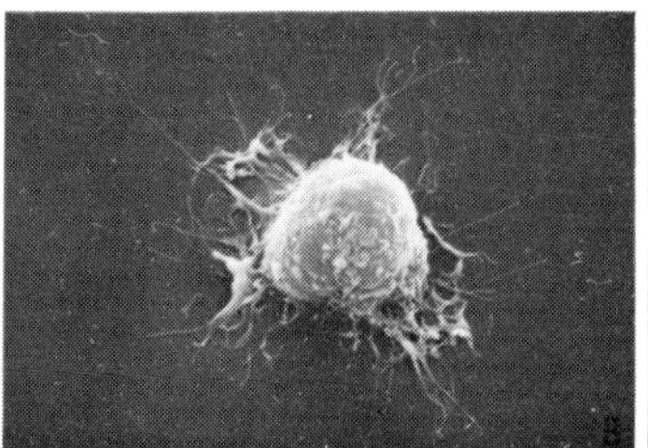
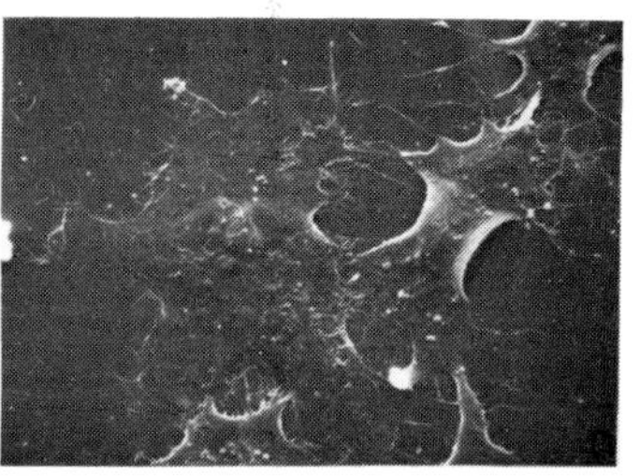

Fig. 6 : Scanning Electron Micrographs of MG 63 attached during 60 min on α_2M (A), α_2M-T (B) or Fibronectin (C).

From these results it can be concluded that α_2M, complexed with a protease allows the spreading of the two cell types. Spreading is more advanced than on Albumin, but slower than on Fibronectin.
Thus the presence of a receptor has not selectively favoured the spreading behavior of the cells which carry the receptor. This again is similar to what has been described for Fibronectin.

CONCLUSION

Serum α_2Macroglobulin complexed with trypsin functions as an attachment factor for receptor-carrying (NHF) and for 'low receptor' (MG 63) cells.
Spreading of both cell types occurred on a coat of α_2M-T but less readily than on a coat of Fibronectin.
Competitive inhibition of attachment or spreading could not be obtained with α_2M-T, Fibronectin, Insulin or EGF in the culture medium.
Thus no evidence was obtained that α_2M-T, although an attachment factor, favoured or altered the attachment of receptor carrying cells as compared to low receptor cells.

REFERENCES

Brugmans, M., J.J. Cassiman, and H. Van den Berghe (1978) J. Cell Sci., 33, 121-132.
Grinnell, F., and D. Minter (1978) Proc. Natl. Acad. Sci. USA 75, 4408-4412.
Heremans, H., A. Billiau, J.J. Cassiman, J.C. Mulier, and P. De Somer (1978) Oncology 35, 246-252.
Rajaraman, R., J.M. MacSween, and R.A. Fox (1978) J. theor. Biol. 74, 177-201.
Seglen, P.O., and J. Fossa (1978a) Exp. Cell Res. 116, 199-206.
Seglen, P.O., and R. Gjessing, (1978b) J. Cell Sci. 34, 117-131.
Van Leuven, F., J.J. Cassiman, and H. Van den Berghe (1978) Exp. Cell Res. 117, 273-282.
Van Leuven, F., J.J. Cassiman, and H. Van den Berghe (1979) J. Biol. Chem. 254, 5155-5160.
Wolpert, L., I. Macpherson, and I. Todd (1969) Nature 223, 512-513.
Yamada, K.M., and K. Olden (1978) Nature 275, 179-184.

Receptor-mediated Uptake of α_2Macroglobulin-protease Complexes by Cultured Cells

F. Van Leuven and J. J. Cassiman

Division of Human Genetics, University of Leuven, Minderbroedersstraat 12, B-3000 Leuven, Belgium

ABSTRACT

Serum components, present intracellularly in cultured human skin fibroblasts, were identified as α_2Macroglobulin (α_2M), albumin, α_1-trypsin inhibitor, hemopexin and transferrin, among others. These components were shown to be taken up from the culture medium. In steady state culture conditions, α_2M was quantitatively the most important serum component taken up by the cells. Complexation of α_2M with trypsin or collagenase largely stimulated the uptake by the cells. The kinetic parameters of initial uptake were examined for native and trypsin complexed α_2M by immunological methods and by ^{125}I-labeling respectively with complete serum and with purified α_2M. Both approaches gave comparable results. The uptake mechanism was of high affinity for α_2M-protease complexes (half maximal uptake reached at 6 x 10^{-8} M α_2M) with a high rate of internalization (about 10^6 molecules of α_2M per cell and per hour under saturating conditions of 200 μg α_2M-trypsin per ml medium). Measurements of binding at 4° C demonstrated the presence of a limited number of surface receptors for α_2M-protease complexes (about 15.000 per cells at 4° C) with an apparent association constant of 0.54 nM^{-1}. Kinetic analysis of uptake and binding data indicate that the effects with native α_2M can be explained by a fraction of complexed and denaturated α_2M always present in purified preparations.
Intracellularly, α_2M becomes localized in the lysosomes, where it is rapidly degraded (T1/2 = 2 hours). Virus-transformed and tumor-derived cell lines showed low or undetectable levels of intracellular α_2M.
Measurement of initial uptake confirmed that these cell lines were unable to take up α_2M-protease complexes. The absence of specific binding at 4° C of ^{125}I α_2M-trypsin to these cells led us to conclude that these cells were devoid of the specific receptors at the cell membrane.

KEYWORDS

α_2Macroglobulin; protease; receptor; fibroblast; tumor cells.

REFERENCES

Van Leuven, F., R. Verbruggen, J.J. Cassiman, and H. Van den Berghe (1977)

Exp. Cell Res., 109, 468-471.
Van Leuven, F., J.J. Cassiman, and H. Van den Berghe (1978) J. Immunol. Methods, 23, 109-116.
Van Leuven, F., J.J. Cassiman, and H. Van den Berghe (1978) Exp. Cell Res., 117, 273-282.
Van Leuven, F., J.J. Cassiman, and H. Van den Berghe (1979) J. Biol. Chem., 254, 5155-5160.

Inhibition of Plasminogen Activator by Sera from Tumor-bearing Mice

B. Nagy

Central Institute for Tumors and Allied Diseases, 41000 Zagreb, Ilica 197, Yugoslavia

ABSTRACT

We studied the production of plasminogen activator in two murine tumors: a spontaneously-developed mammary carcinoma and a methylcholanthrene-induced fibrosarcoma in C3Hf/Bu mice. Mammary carcinoma is a highly metastasizing tumor in contrast to a fibrosarcoma which is low. Fibrinolysis was assayed on ^{125}I-fibrin-coated Petri dishes employing a reaction between tumor cell lysates and plasminogen. Both tumors contained high levels of plasminogen activator as indicated by the increase of fibrinolysis. This fibrinolytic activity of tumor cell extracts can be inhibited by sera from either normal or tumor-bearing mice. However, sera from tumor-bearing mice were more inhibitory than normal, but their activity increased as the tumor size enlarged.

KEYWORDS

Plasminogen activator; fibrinolysis; serum protease inhibitors; tumor growth; mouse fibrosarcoma; mouse mammary carcinoma.

INTRODUCTION

It has been reported that mammalian fibroblast cultures transformed by oncogenic tumor viruses or primary cell cultures from spontaneous or artificially-induced animal tumors exhibit increased fibrinolytic activity when compared to corresponding normal cell cultures (Unkeless, 1973; Ossowski, 1973). More recently (Nagy, 1977), increased fibrinolytic activity has also been detected in biopsies from human tumors. Fibrinolytic activity is generated by serine protease, an enzyme arginine-specific protease of urokinase type, that acts as a plasminogen activator; this factor converts plasminogen into plasmin which then degrades locally accessible proteins (Unkeless, 1974).
Fibrinolytic activities are correlated with the expression of a number of phenotypic

characteristics of cell transformation (Ossowski, 1975). Plasminogen activator production, however, is not always indicative of the transformed state (Lange, 1975; Laishes, 1976) but it is characteristic of invasive and migratory cells (Sherman, 1976). It could be assumed then that the inhibition of this enzyme may constitute an important natural defence mechanism against tumor invasiveness. Fibrinolysis mediated-plasmin can be neutralized by several plasma protease inhibitors such as α_2-macroglobulin, α_1-antitrypsin, esterase inhibitors and inter-α-trypsin inhibitor. Moroi (1976) has recently reported the isolation of an α_2-globulin with both antiplasmin and antiactivator activities. A new plasmin inhibitor, named antiplasmin was also identified in human plasma (Collen, 1977). These inhibitors (Moroi, 1976; Collen, 1977) differ from the plasma protease inhibitors mentioned above as well as from the inhibitor described by Hedner (1976). Loskutoff (1978) found a reduction in the fibrinolytic acitivity of transformed cells when foetal bovine serum was employed as a source of plasminogen, and this reaction ascribed to inhibitor substances present in the serum. It appears, therefore, that there is more than one fibrinolytic inhibitor present in the plasma which acts by inhibiting plasmin and/or the plasminogen activator associated with oncogenic transformation or neoplasia.

Increased fibrinolytic activity may be responsible for many of the altered morphological and behavioral properties of tumors. Also, factors which regulate fibrinolytic activity may affect the expression of the transformed phenotype.

In the studies described here, we investigated whether two murine malignant tumors, a spontaneously developed mammary carcinoma and a methylcholanthrene-induced fibrosarcoma, exhibit fibrinolytic activity and whether this activity is related to the ability of the tumors to produce metastases.

MATERIAL AND METHODS

Mice and Tumors

Female C3Hf/Bu mice from the conventional breeding colony were used. The fibrosarcoma was originally induced in a young female C3H mouse by a single subcutaneous injection of 1 mg 3-methylcholanthrene suspended in peanut oil (Suit, 1967), and the mammary carcinoma developed spontaneously in an old female mouse of the same strain (Milas, 1975). While fibrosarcoma rarely metastasizes, mammary carcinoma gives rise to metastases in the lung regularly. Single cell suspensions from fibrosarcoma and mammary carcinoma were prepared by a mechanical method (Milas, 1975). Tumors in the leg were generated by injecting 2-5 x 10^5 viable tumor cells into the muscles of the right thigh. All mice developed tumors within 10 days. Tumor growth curves were obtained by measuring three mutually orthogonal diameters of tumors and then calculating the mean values. At different time intervals the mice were sacrified and their sera were prepared for the assay (see below). Tumors were also harvested, the tissue was minced and cell lysates were prepared.

Preparation of Cell Lysates containing Plasminogen Activator

Tumor pieces were cleaned of necrotic and hemorrhagic regions and then mechanically despersed into tiny pieces. These were washed several times with buffer and then minced and teased until a cell suspension was obtained. The suspension was centrifuged at 300 g, the sediment was resuspended in phosphate-buffer saline (0.15 M NaCl; 0.005 M sodium phosphate, pH 7.8) and subjected to homogenisation in a homogenizer with a tight fitting pestle using 10-20 strokes. An equal volume of 0.5% Triton X-100 was added and the treatment with homogenizer was repeated, followed by centrifugation at 300-500 g for 10 minutes. The nuclear pellet was discarded and the supernatant was assayed for fibrinolytic activity.

Preparation of ^{125}I-fibrinogen and Plasminogen

Bovine fibrinogen was partially purified by precipitation (Laki, 1951) and iodinated (Helmkamp, 1960). The iodination procedure was slightly modified: cold 0.0005 M ICl was added to 5 mCi of ^{125}INa and this was rapidly mixed with a solution of purified fibrinogen (10 mg) in 1 ml of borate buffer pH 8.0 (0.16 M NaCl; 0.20 M H_3BO_3). The reaction mixture was incubated for 8 min at room temperature and after passage over a column of Dowex-1-Cl^- (0.6 x 2.5 cm) dialyzed against phosphate-buffer saline (PBS). Plasminogen was isolated from plasma by affinity chromatography on a lysine-sepharose column as described by Deutsch (1970).

Assays

The assay for plasminogen-dependent fibrinolysis was based on release of ^{125}I-labelled fibrinopeptides from ^{125}I-fibrin-coated Petri dishes (Unkeless, 1973). Petri dishes (35 mm Ø) were coated with 0.1 mg ^{125}I-fibrinogen (final radioactivity 50,000 - 100,000 counts/min) dissolved in 0.1 ml phosphate-buffered saline (PBS). The Petri dishes were dried for 24 h at 45°C. After that the fibrinogen was converted to fibrin by incubating the plates with MEM culture medium containing 10% foetal calf serum for 2 h at 37° C followed by thorough washing with cold TD buffer. To assay the fibrinolytic activity of cell lysates, the reaction mixture (in final volume 1 ml) contained 0.1 M Tris pH 8.1, 4 μg of human plasminogen, and cell-free supernatant of desired protein concentration. This was added to ^{125}I-fibrin coated Petri dishes and incubated at 37°C for 4 hours. To test the inhibitory effect of mouse serum, assays were performed in Petri dishes containing ^{125}I-fibrin. Tumor cell lysates and murine sera were then added to serve as a source of plasminogen activator and plasminogen, respectively. One half of the Petri dishes contained untreated serum and another half contained serum previously treated with 1 M HCl at pH 3 for 2 hours.

RESULTS

Fibrinolytic activity of mammary carcinoma cell lysates increased with the increase in protein concentration, Fig. 1. The activity was already high ($\sim$35% of total activity) at the concentration of 10 μg of protein. It progressively increased and at the protein concentration of 100 μg, was more than 70% of the total activity. Similar activity was obtained with lysates from fibrosarcoma cells, Fig. 2. In this case 4 μg of pure plasminogen was added to the Petri dishes instead of acid treated tumor-bearing mouse serum.

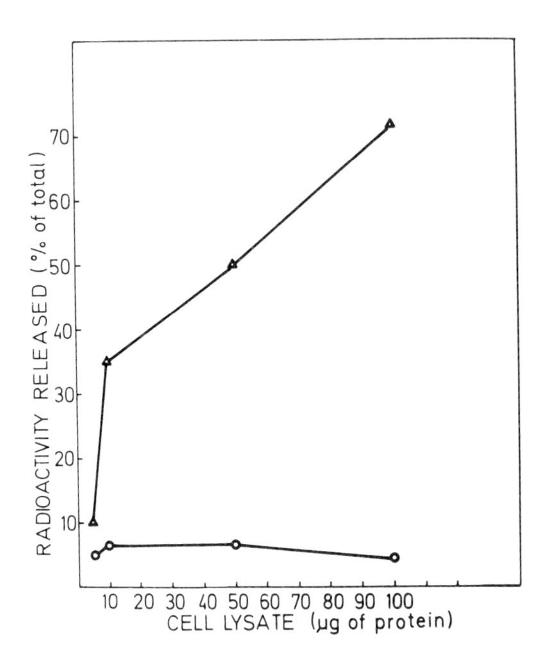

Fig. 1. Fibrinolytic assay of mammary carcinoma cell lysate as a function of protein concentration. The indicated protein amount was added (in a final volume 1 ml) to 0.1 M Tris-HCl, pH 8.1 and the serum concentration was 3%. Circles - serum of mice bearing mammary carcinoma, triangles - acid treated serum of mice bearing mammary carcinoma.

The following experiment was designed to study whether normal mouse sera, healthy human sera or sera from mice bearing mammary carcinoma or fibrosarcoma inhibit fibrinolytic acitivity of fibrosarcoma cell lysates. Different concentrations of acid untreated sera, ranging from 0.5 to 10%, were added to 100 µg of cell lysates (protein). Figure 3 and 4 indicate that all sera used inhibited the fibrinolysis. This inhibition was more pronounced with higher serum concentrations. In addition, sera from tumor-bearing mice were more inhibitory than serum from normal mice, Fig. 4.
We next investigated the time appearance of the increased inhibitory effects of sera from tumor-bearing mice upon the fibrinolytic activity.
Sera were derived from mice 4, 8, 13, 18, 21 or 25 days following subcutaneous transplantation of 4×10^5 mammary carcinoma cells. The results presented in Fig. 5 show that a significant increase in the activity appeared at day 13 following tumor cell transplantation, at the time when tumors grew to about 6 mm in

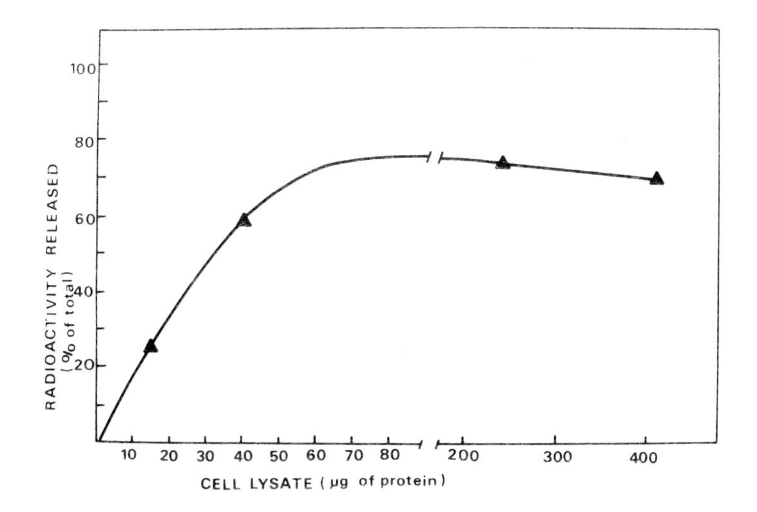

Fig. 2 Fibrinolytic assay of fibrosarcoma cell lysate as a function of protein concentration. The indicated protein amount was added to 0.1 M Tris-HCl, pH 8.1 and 4 μg of plasminogen.

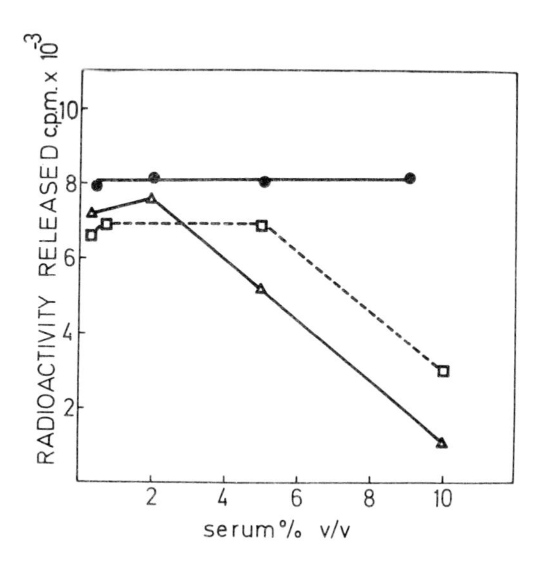

Fig. 3 Effect of serum concentration on fibrinolytic activity of fibrosarcoma cell lysate. Dark circles - 4 μg plasminogen, squares - mouse serum and triangles - human serum. Serum was added to 0.1 M Tris-HCl, pH 8.1 and 100 μg of cell lysate protein.

diameter. After that, the inhibitory activity of sera remained approximately at the same level.

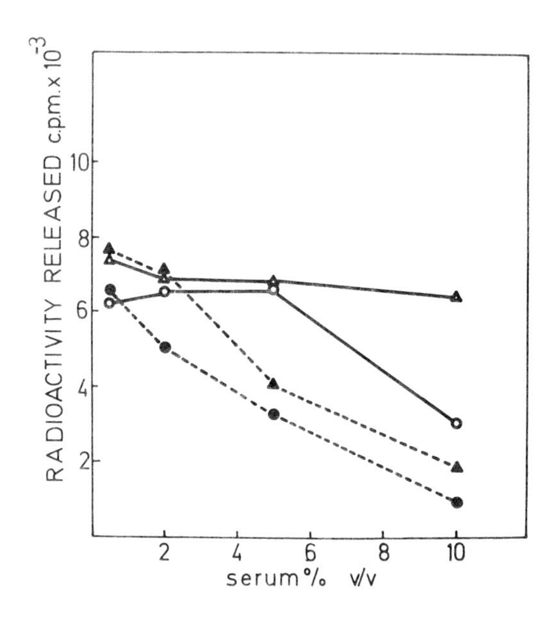

Fig. 4 Effect of serum concentration on fibrinolytic activity of fibrosarcoma cell lysate. The indicated serum amount was added to 0.1 M Tris-HCl, pH 8.1 (in a final volume 1 ml) and 100 μg of cell lysate protein. Plain circles - normal mouse serum, plain triangles - 4 μg of plasminogen, dark circles - mouse bearing mammary carcinoma serum and dark triangles - serum of mouse bearing fibrosarcoma + 4 μg of plasminogen.

DISCUSSION

As already mentioned in the Introduction, in vitro transformed cells (Ossowski, 1973; Unkeless, 1973) and cells of many malignant tumors (Nagy, 1977) contain high levels of fibrinolytic activity which is ascribed to the increased production of the plasminogen activator by these cells. This is not the exclusive property of malignant cells since certain normal tissues also produce the plasminogen activator (Lang, 1975; Laishes, 1976). Although the role of this factor has been implicated in many aberrant properties of neoplastic cells, its association with tumor invasion and metastases has not been fully investigated. There are studies which imply that increased fibrinolysis is in direct correlation with the capacity of tumors to infiltrate surrounding tissue and produce metastases (Ambrus, 1975). In our experiments, which we described here, we determined whether there is any difference

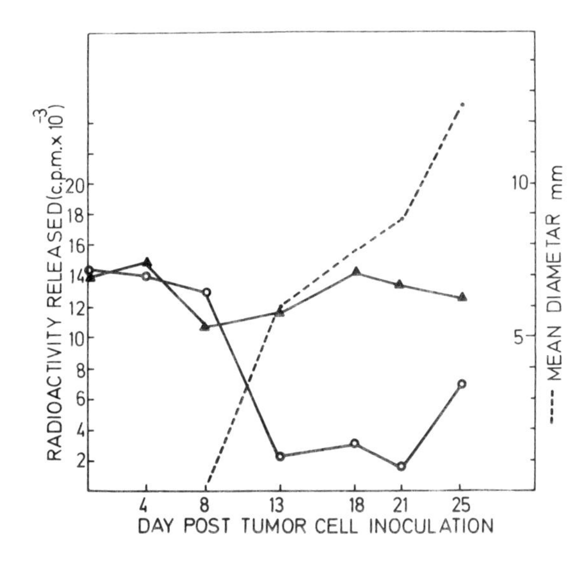

Fig. 5 Fibrinolytic activity inhibition in the serum of mice bearing mammary carcinoma. After 4 x 10^5 mammary carcinoma cells were injected into the muscles of the right thigh, the tumor size was measured (dashed line) and the mice were killed at the indicated time. Mice serum was harvested and divided into two. One half of the serum was added to the reaction mixture (plain circles), since the other part of each serum was treated with pH-3 for 2 h to inactivate the inhibitors (dark triangles). The reaction mixture consisted of 100 ug of protein cell lysate in 1 ml of 0.1 M Tris-HCl, pH 8.1.

in the content of plasminogen activator between two murine tumors which greatly vary in their capacity to metastasize. The results indicated that both fibrosarcoma and mammary carcinoma (low and high metastazing tumor, respectively) show similar fibrinolytic activities as expressed per /ug of isolated protein. Both tumors also cause a similar elevation in the inhibitors of the plasminogen activator present in the serum. According to data obtained with a mammary carcinoma, an increase in the serum's inhibitory activity was not present soon after tumor cell transplantation. The activity appeared when tumors achieved a certain size (approximately 6 mm in diameter) and remained at that level in spite of further tumor progression.

In conclusion, our results indicate that the plasminogen activator activity of cells derived from a highly metastasizing mammary carcinoma was similar to that of cells derived from a fibrosarcoma with low capacity to produce metastases. Also, production of plasminogen activator by cells of both tumors is followed by the appearance of increased activity of sera to inhibit fibrinolytic activity of this factor.

ACKNOWLEDGEMENT

The author wishes to express her thanks to Professor L. Milas for his support and valuable comments on the manuscript. The technical assistance to Mrs. Marija Tajber and Mrs. Ada Lechpammer is gratefully acknowledged.

REFERENCES

Ambrus, J.L., C.M. Ambrus, J. Pickern, S. Soldes, and I. Bross (1975). J. Med. 6, 433-458.

Collen, D., A. Billian, J. Edy, and P. De Somer (1977). Biochim. Bioph. Acta,499, 194-201.

Deutsch, D.G., and E.T. Mertz (1970). Science, 170, 1095-1096.

Hedner, U., and D. Collen (1976). Thromb. Res. 8, 875-879.

Helmkamp, R.W., R.L. Goodland, W.F. Bale, J.L. Spar, and L.E. Mutschler (1960). Cancer Res., 20, 1495-1500.

Laishes, B.A., E. Roberts, and C. Burrowes (1976). Biochem. Biophys. Res. Commun., 72, 462-471.

Laki, K. (1951). Arch. Biochem. Biophys., 32, 317-324.

Lang, W.E., P.A. Jones, and W.F. Benedict (1975). J. Natl. Cancer Inst., 54, 173-176.

Loskutoff, D.J. (1978). J. Cell. Physiol., 96, 361-370.

Milas, L., N. Hunter, I. Bašić, K. Masan, D.J. Grdina, and H.R. Withers (1975). J. Nat. Cancer Inst., 54, 895-902.

Moroi, M., and N. Aoki (1976). J. Biol. Chem.,251, 5956-5965.

Nagy, B., J. Ban, and B. Brdar (1977). Int. J. Cancer, 19, 614-620.

Ossowski, L., J.P. Quigley, and E. Reich (1975). In: Proteases and Biological Control, (Reich, E., Rifkin, D.B., and Shaw, E., eds.), p. 901.

Ossowski, L., J.C. Unkeless, A. Tobia, J.P. Quigley, D.B. Rifkin, and E. Reich (1973). J. Exp. Med., 137, 112-126.

Sherman, M.I., S. Stricland, and E. Reich (1976). Cancer Res., 36, 4208-4216.

Suit, H.D., and D. Suchato (1967). Radiology, 89, 713-719.

Unkeless, J.C., K. Danø, M. Kellerman, and E. Reich (1974). J. Biol. Chem., 249, 4295-4305.

Unkeless, J.C., A. Tobia, L. Ossowski, J.P. Quigley, D.B. Rifkin, and E. Reich (1973). J. Exp. Med., 137, 85-111.

Tubulin Localisation in Whole, Glutaraldehyde Fixed Cells, viewed with Stereo High-voltage Electron Microscopy

J. De Mey, J. J. Wolosewick*, M. De Brabander, G. Geuens, M. Joniau and K. R. Porter***

Laboratory of Oncology, Janssen Pharmaceutica Research Laboratories, B-2340, Beerse, Belgium
**MCD Biology, University of Colorado, Boulder, Colorado 80309, U.S.A.*
***Interdisciplinary Research Center, Catholic University of Louvain, Campus Kortrijk, B-8500 Kortrijk, Belgium*

ABSTRACT

The intracellular distribution of microtubules and cytoplasmic tubulin has been studied with a new technique for the high resolution three-dimensional localisation of intra- and extracellular antigens. This approach combines the unlabeled Peroxidase-anti-peroxidase method with high-voltage electron microscopy of whole, critical-point-dried cells. It greatly increases the resolving power for observing whole, antibody-stained cells. When antibodies to tubulin are used the following observations can be done:

1) Stained threads which appear single at the light microscopic level are frequently resolved as bundles of 2-3 individual microtubules when viewed at the E. M. level. This indicates that MT bundling may be a more general phenomenon than one would expect from observations on thin sections

2) The label is associated with the microtrabecular lattice of the cytoplasmic ground substance, but in a non-uniform way. Large regions of it are not or only lightly stained. However, a consistent and uniform concentration of label is found on the trabeculae associated with the microtubules. If these observations reflect the actual *in vivo* situation, they show a less random distribution of the tubulin pool than one would deduce from thin sections. It is our hope that this new approach will provide more accurate information on the molecular mechanisms involved in the *in vivo* microtubule polymerisation-depolimerisation, and their role in cell locomotion, intracellular movement and cell surface and shape modulations.

KEYWORDS

Microtubules, tubulin, immunoelectron microscopy, whole mount cells, high voltage electron microscopy, cytoplasmic ground substance.

INTRODUCTION

De Brabander and co-workers (1976, 1977c), and also Vasiliev and Gelfand (1976) and Bhisey and Freed (1971) have given clear evidence that microtubules are playing an important role in the cell's capacity of directional locomotion. Also from the work of Mareel and Marc De Brabander (1978) one can learn that cancer cells cannot invade normal tissues when their cytoplasmic microtubular complex is destroyed by antimicrotubular drugs. A better understanding of the basic mechanisms involved in microtubule functioning, and their role in cell locomotion, intracellular movement and cell surface and cell shape modulations is therefore a goal set forward by many investigators. Until some years ago, thin sectioning electron microscopy was the only way to study the structure and intracellular distribution of microtubules (MT), microfilaments and intermediate filaments. The recent adaptation of immunofluorescence microscopy by Lazarides and Weber (1974) and Weber and co-workers (1975), using specific antibodies to the proteins contained within their structure has given valuable information on the gross distribution of these proteins in different cell lines.
In order to obtain a method, also useful for ultrastructural studies, we have explored an immunoperoxidase approach for the visualization of the MT system, and this after adequate fixation with glutaraldehyde (De Brabander and co-workers, 1977a, 1977b, 1979; De Mey and co-workers, 1976, 1978). In spite of its many advantages, this approach, however, has still a number of limitations: the resolution for the study of whole cells at the light microscopic level is not better than with immunofluorescence. In addition, thin sectioning of embedded cells results in a loss of much essential information on three-dimensional relationships. This makes interpretations difficult. To achieve a more complete approach, we wanted to explore the possibility of combining the recently introduced use of high voltage electron microscopy (HVEM) on whole, critical-point-dried (CPD) cells, (Buckley, 1975; Buckley and Porter , 1975, Gershenbaum, Shay and Porter, 1974; Wolosewick and Porter, 1976), with our immunoperoxidase method. With the help of stereotechniques, it is possible to study with this relatively simple procedure the three-dimensional intracellular relationships between the different cell organelles and fibrillar systems. The method has thus far lead to discovery of a space filling lattice of fine filaments that is believed to represent the native confirmation of the cytoplasmic ground substance in which the different organelles and fibrillar systems are suspended. By combining this new approach in morphology with the PAP-immunocytochemical method, we hoped to better define the distribution of specific proteins, and in this paper, we present results obtained with specific antibodies to tubulin.

MATERIALS AND METHODS

Cell Culture

Growth conditions for mouse MO and rat kangaroo Ptk_2 cells used in this work were as described by De Brabander and co-workers (1976). For the experiments, the cells were seeded on formwar-coated (0.7 % in chloroform) 100 mesh golden grids in the following way. 1 ml of complete medium was added to 30 mm diameter plastic petri dishes. The desired number of sterilized coated grids were than placed on the bottom of the dish, formwar side up. To this set up, 1 ml of cells in medium were seeded at different densities and allowed to grow for at least 24 h.

Fixation and permeabilization

Cells were fixed and rendered permeable to antibodies according to the method recently described by Weber, Rathke and Osborn (1978). Care was taken to avoid air drying during changes of solutions.

Immunocytochemistry

Immunocytochemistry and appropriate controls were done as described by De Brabander and co-workers (1977), but with a few modifications:
1. Normal goat serum (NGS) 1/20: 30 min. 2. Antibody to tubulin (Joniau and co-workers, 1977), 0.25 µg/ml in 1 % NGS/TBS overnight at 4° C. 3. GAR/Ig (GIBCO), 1/20 in TBS, 1 h room temperature. 4. PAP, 1/1000 in 1 % NGS/TBS, 1 h. Between steps 2 and 3, steps 3 and 4 and after step 4, the preparations were washed with TBS (10 m M Tris in saline, pH 7.6) for 3 x 10 min on a rocker platform. The peroxidase reaction was carried out at pH 7.0. Cells were reacted with 1.0 % OSO_4 in bidistilled H_2O for 10 min, dehydrated through a series of acetone, critical-point-dried from CO_2 and carbon coated at both sides. In order to study morphological preservation, cells were grown on formwar-coated grids, fixed and permeabilized in the same way, but directly osmicated, dehydrated and processed for CPD. In addition, some preparations were fixed according to standard procedures and processed for CPD according to Wolosewick and Porter (1976). All the experiments done on grids were also done on cells on cover glasses. They were processed for thin sectioning as described by De Brabander and co-workers (1977). The cells were observed with a Leitz-Orthoplan microscope, a Philips 201 C, operated at 100 KeV, and JEOL-1000 and - 1250 HVEM's at 1000 KeV, all equipped with a tilting stage.

RESULTS

At the light microscopic (LM) level, cells stained with tubulin antibodies display the now familiar picture of the cytoplasmic microtubular complex. Control preparations can be observed only with phase optics, due to the virtually complete absence of reaction product (De Brabander and co-workers, 1977a).

At the EM level, the ultrastructural preservation observed in thin sections is very acceptable and slightly improved in comparison with our previous method (De Brabander and co-workers, 1977a). The ultrastructural preservation of the whole mount cells can be evaluated best in the control preparations (i.e., with the omission of tubulin-antibody). They are completely negative with regard to peroxidase reaction product which confirms the high degree of specificity of the PAP-procedure. The pictures obtained are similar to

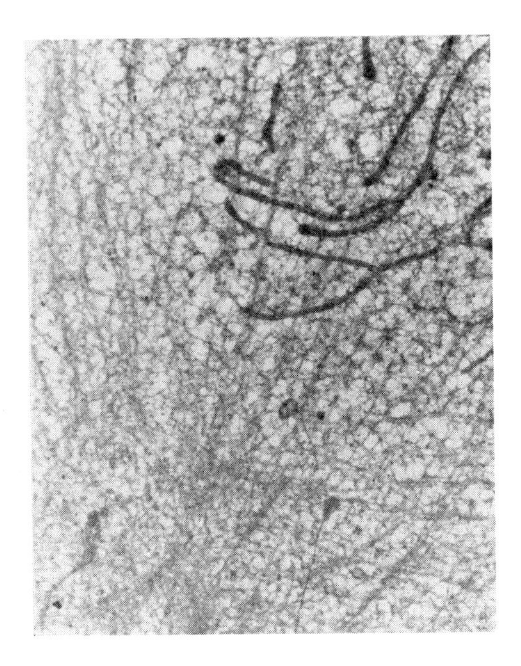

Fig. 1: Control preparation of a Ptk_2 cell showing acceptable preservation and absence of any reaction product. x 6000

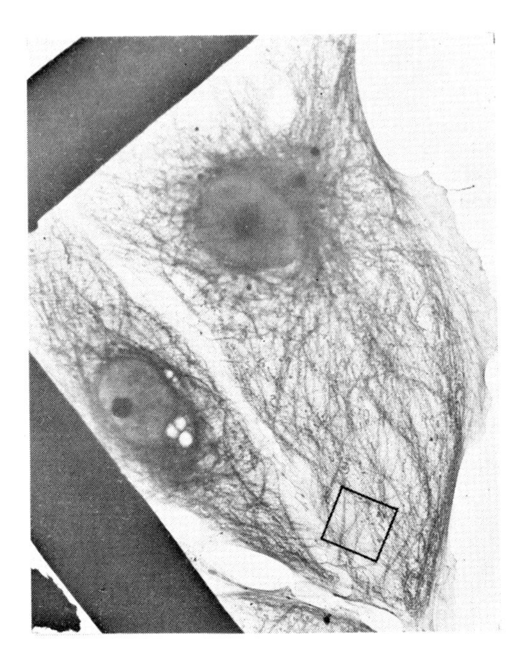

Fig. 2: Low magnification of a Ptk_2 cell stained with tubulin antibodies. The resolution is similar to that of the light microscope. The region within the rectangle is shown in Fig. 3. x 720

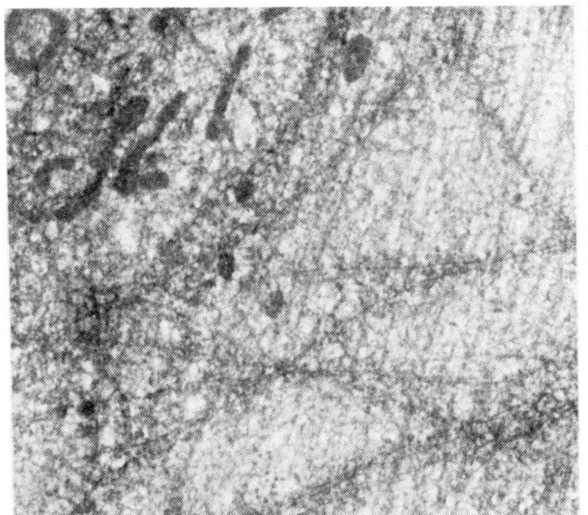

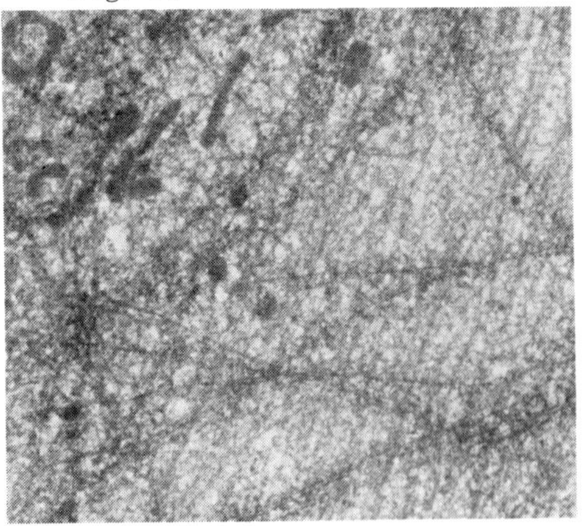

Fig. 3. A stereo picture of part of Fig. 2. The labeled strands of Fig. 2 can now be identified as MT's. Note the dramatic increase in resolution, and the occurrence of numerous small MT bundles which are not visible in Fig. 2. Label is also seen on the trabeculae surrounding and attaching onto MT's. x 6000

those obtained with conventionally fixed preparations. The different subcellular organelles such as microfilaments, stress fibers, mitochondria, lysosomes, endoplasmic reticulum and MT are easily recognizable. Microtubules, although showing considerable shrinkage, can be traced over long distances. Discontinuities (or breaks) are not noted. In particular the so-called microtrabecular lattice has retained its dimensions, spatial organization and association with the different subcellular structures, including MT's (Fig. 1).

Cells stained with tubulin antibodies and viewed at low magnification in the HVEM give a picture nearly comparable with the light microscope one (Fig. 2). Due to their uniform labeling MT's can be followed over very long distances, except in regions where the cell becomes too thick. At increasing magnifications in the HVEM, an apparently single thread often is resolved into two to three labeled strands which can be identified as single MT's (Fig. 3).

At still higher magnifications, and particularly when viewed in stereo, it is observed that almost all MT's are covered with reaction product in a way typical for the PAP-method (De Mey and co-workers, 1976).

Additionally, cytoplasmic label can be seen. It is always associated with the microtrabecular lattice and with the considerable number of contained ribosomes (Fig. 3). The distribution of this cytoplasmic label is, however, not uniform.

Large parts of the lattice are only lightly stained, especially in the thinner margins of the cells. A consistent and uniform amount of label is found also on the trabeculae surrounding and attaching to MT's. This is best visible around single, widely separated MT's, and leads to production of very complex pictures in MT bundles. Besides MT and cytoplasmic label, several membrane-limited organelles are consistenly covered with large amounts of label, which is shown to be immunospecific by the appropriate control experiments. Among them are the mitochondria, cisternae of the endoplasmic reticulum, lysosomes and vacuoles. Microfilaments, stress fibers and intermediate filaments are not labeled.

DISCUSSION

Using a combination of LM and thin section EM we have previously published the first data on the probable distribution of non-microtubular tubulin (De Brabander and co-workers, 1977a, 1977b, 1979). In order to achieve a more complete approach, we have developed a method combining the capabilities of the HVEM on whole CPD cells and PAP-immunoperoxidase method for the localisation of intracellular antigens. The fixation and permeabilization procedure gives quite acceptable ultrastructural preservation. Apparently, the method permits adequate antibody penetration, while giving reasonable preservation of both ultrastructural detail and antigenicity.

Limitations of the method

Our present observations provide a fair impression of the new possibilities of the technique, but also suggest its limitations. With regard to the latter, it should be stressed that the procedure has the limitations inherent to all immunocytochemical methods, as has been discussed before (De Brabander and co-workers, 1977, 1979). In spite of the penetration offered by the higher-energy electron beam, the method is still limited to the thinner parts of the cells, because the images of the thicker parts contain more structural information than one can unscramble. Here, HVEM of semi-thin (0.5 μ) sections, will probably be very useful as an alternative. Another limitation is inherent in the rather poor definition of the reaction product of peroxidase. Other markers such as ferritin and colloïdal gold are currently being tested. In addition, micrographs of whole cells seldom give the fine structural detail obtained by thin sectioning. Therefore, the best approach for the exact localisation of an antigen will certainly be a combination of all possible approaches, including observations on conventionally treated preparations.

New possibilities of the method

The gain of information offered by the technique is illustrated by our observations on cells stained with antibodies to tubulin. The gap between the light microscope visualization of whole MT networks and EM fine distribution studies in thin sections has been filled completely. Both can now easily be investigated in the same cells and in virtually unlimited number of cells. When the cells are observed at middle high magnifications, it is often seen that apparently single threads are resolved into 2 to 3 individual MT's, forming a small bundle. While it is true that at the LM level under optimal conditions of immunofluorescence, single MT's can be seen (Osborn , Webster and Weber, 1978), one can never at this level of resolution be absolutely sure that a stained thread is in fact a single MT, rather than a small bundle. Our present results show that the small bundles are very common in the cells we studied. Lateral association of MT may be a more general phenomenon than one would expect from observations on thin sections. This may be of considerable importance in understanding the mechanisms by which the cytoplasmic microtubule complex performs its various functions. In this connections, it should be interesting to see whether the degree of bundling is similar in various cell types (e.g. epitheloid versus fibroblastic, and/or transformed versus not-transformed).

From our data on the probable distribution of non-microtubular tubulin, we concluded that this tubulin is excluded from the interior of membrane lined organelles, and evenly distributed throughout the cytoplasm. Microfilaments and intermediate filaments are unstained, but label is often seen at the cytoplasmic surface of organelles such as mitochondria, E.R., lysosomes and the nucleus. Most of these observations could be confirmed by the present method. However, the observations on the distribution throughout the cytoplasm can now be refined and partly corrected by the stereo, three-dimensional approach. Indeed, the cytoplasmic label is seen to be associated with the microtrabecular lattice and with large numbers of ribosomes contained within it. The staining of the lattice is however not uniform. Major parts of it are not or only lightly stained, especially in thin peripheral lamellae. On the other hand, a consistant and uniform concentration of label is found on the trabeculae closely associated with MT's. These observations, if they reflect the actual in vivo situation, therefore show a much less random distribution of the tubulin pool than could be deduced from thin sections. The possible implications with regard to MT turnover and functioning are obvious and must not be discussed here. These corrections should indeed be certified by further experimentation. In particular, the following possible objections should be kept in mind: i) displacement of the antigen by the procedure; ii) diffusion of reaction product from zones of high antigen concentration; iii) minor contamination of the affinity purified antibody solution by unrelated antibodies (e.g. MAPS), although they could not be detected by standard immunochemical procedures (Joniau and co-workers, 1977).

In conclusion, we have presented results showing the feasability of combining immunocytochemistry and HVEM on whole, adequately-fixed and CPD cells. This new approach in morphology will provide us with new and complementary possibilities by dramatically increasing the resolving power of immunotechniques in the study of whole cells.

Some points, however, still need to be clarified and the method will need some refinement, under the form of a better defined label. For example, recently, Webster and co-workers (1978) succeeded in using an indirect immunoferritin method for labeling MT's and actin containing filaments in isolated detergent extracted "cytoskeletons". It remains uncertain whether ferritin will be useful in HVEM. In spite of this, we are able now to present a number of new findings, indicating that it will be possible to obtain relevant data on *in vivo* mechanisms of MT turnover and functioning. We are currently studying the breakdown of MT's by anti MT-drugs and their reformation after washing the drug away. The method is applicable to other antigens as well and may therefore give a new impetus to the study of the other filamentous systems, and especially in the elucidation of the nature of the microtrabecular system.

ACKNOWLEDGEMENT

We gratefully acknowledge Prof. Dr. Amelincks and Prof. Dr. Van Landuyt from the Centre for High Voltage Electron Microscopy at the RUCA, Antwerp, for the use of the High Voltage Microscope installed there. This work was supported by grants from the Instituut voor Bevordering van het Wetenschappelijk Onderzoek in de Nijverheid en Landbouw (Brussels, Belgium), and from the Nationaal Fonds voor Wetenschappelijk Onderzoek (Brussels, Belgium). J. D. M. acknowledges a grant from the Belgian N. F. W. O. for his journey to the MCD Biology department at the University of Colorado in Boulder and also Prof. Dr. W. Gepts for facilities in the laboratory of Experimental Pathology, Free University of Brussels, where he was affiliated during the early stages of development of the technique. The research carried out in the Department of Molecular, Cellular and Developmental Biology in Boulder was supported by NIH Grant # 5P41 RR00592-09. The authors are especially grateful to Judy Fleming for her expert technical assistance.

REFERENCES

Bhisey, A. N. and J. J. Freed (1971). *Exp. cell. Res., 64,* 419-429.

Buckley, I. (1975). *Tissue and Cell., 7,* 51-72.

Buckley, I. K. and K. R. Porter (1975). *Journ. of Micros., 104,* 107-120.

De Brabander, M., R. Van de Veire, F. Aerts, M. Borgers and P. A. J. Janssen (1976). *Cancer Res., 36,* 905-916

De Brabander, M., J. De Mey, M. Joniau and G. Geuens (1977a). *J. Cell. Sci., 28,* 283-301.

De Brabander, M., J. De Mey, M. Joniau and G. Geuens (1977b). *Cell Biol. Int. Rep., 1,* 177-183.

De Brabander, M., J. De Mey, R. Van de Veire, F. Aerts and G. Geuens (1977c). *Cell Biol. Int. Rep., 1,* 453-461.

De Brabander, M., G. Geuens, J. De Mey and M. Joniau (1979). Biol. Cell., 34, 213-226.
De Mey, J., J. Hoebeke, M. De Brabander, G. Geuens and M. Joniau (1976). Nature (London), 264, 273-275.
De Mey, J., M. Joniau, M. De Brabander, W. Moens and G. Geuens (1978). Proc. Natl. Acad. Sci., U.S.A., 75, 1339-1343.
Gershenbaum, M.R., J.W. Shay and K.R. Porter (1974). Proc. of Adv. in Biomed. Appl., Research Institute, Chicago, Illinois.
Joniau, M. M. De Brabander, J. De Mey and J. Hoebeke (1977). FEBS letters, 78, 307-312.
Lazarides, E. and K. Weber (1974). Proc.Natl. Acad. Sci., U.S.A., 71, 2268-
Mareel, M.K. and M. De Brabander (1978a). J. Natl. Cancer Inst., 61, 787-792.
Osborn, M, R.E. Webster and K. Weber (1978). J. Cell. Biol., 77, R-27-R-34.
Vasiliev, J.M. and I.M. Gelfand (1976). Cell Motility, ed. Goldman, R., T. Pollard and J. Rosenbaum. Cold Spring Harbor. p. 279-304.
Weber, K., P.C. Rathke, and M. Osborn (1978). Proc. Natl. Acad. Sci. U.S.A. 75, 1820-1824.
Weber, K., A. Pollack and T. Bibring (1975). Proc. Natl. Acad. Sci., 72, 459-463.
Webster, R.E., D. Henderson, M. Osborn and K. Weber (1978). Proc. Natl. Acad. Sci., U.S.A., 75, 5511-5515.
Wolosewick, J.J. and K.A. Porter (1976). Am. J. Anat., 147, 303-324.

The Organized Assembly and Function of the Microtubule System throughout the Cell Cycle

M. De Brabander, G. Geuens, J. De Mey and M. Joniau*

Lab. Oncology, Janssen Pharmaceutica Res. Labs., B-2340 Beerse, Belgium
**University of Leuven, Campus Kortrijk, B-8500 Kortrijk, Belgium*

ABSTRACT

The functional importance of microtubules in cell migration and malignant invasion is shortly reviewed.

The organized assembly of the microtubule apparatus in interphase and mitotic cells is investigated in living cells after release from nocodazole block. The reassembly is followed with peroxidase-anti-peroxidase immunocytochemistry combined with toluidine blue staining of the chromatin and electron microscopy.

The data show that the centrosomes and the kinetochores can indeed nucleate the assembly of microtubules in living cells. Moreover, it is suggested that a mutual interaction is responsible for the preferential elongation of microtubules between these microtubule organizing centres. This can be explained if it is assumed that nucleation is not a strictly local phenomenon but a zonal effect.

KEYWORDS

Microtubule assembly; immunocytochemistry; microtubule organizing centres.

INTRODUCTION

Microtubules in cell movement and invasion

It has long been recognized that many cells and most notably cancer cells are able to migrate along considerable distances. During the last decades the knowledge of the molecules involved in cell motility has made large progress. Two major points seem now to be well established. On the one hand it was found that the force-producing molecules of muscle cells -actin, myosin and associated proteins- are present in large amounts in most animal and plant cells including cancer cells. It is now generally accepted that these molecules constitute the basic machinery used by many cells to produce the mechanical energy necessary for locomotion (for review see Lazarides and Revel, 1979). In the case of cancer cells and other vertebrate cells it is believed that the actomyosin system is involved in the formation of cytoplasmic extensions of varying shape and denomination (filopodia, lamellipodia, lobopodia etc...). When these pseudopodia are able to adhere firmly to a more or less solid substratum, force can be generated in order to move the cell body forward. On the other hand some kind of organizing or coordinating system must exist for an effective use of the force generating machinery, which produces net locomotion or directional migration. The cells must be able to form new extensions in a polarized way. Experiments with microtubule inhibitors (colchicine, vinca alkaloids, nocodazole) have clearly shown that this coordination depends largely on an intact microtubule system. (For review see Dustin, 1978). The destruction of the microtubule system does not interfere with the intrinsic locomotory capacity of the cells. However, the formation of protrusions occurs in a completely disorganized way. As a result the cells loose their polarized aspect and become unable to perform net translocations over the substratum. We could not confirm the generally held belief that microtubules would form an endoskeleton which endows the cell with a polarized shape by providing internal rigidity. As an alterative hypothesis we have proposed that the microtubule apparatus regulates the activity of the actomyosin system in space and time by serving as an intracellular signal transmitting system. (De Brabander and co-authors, 1977).

The essential role of an intact microtubule system for the ability of cancer cells to invade surrounding tissues has recently been demonstrated by Mareel and De Brabander (1978). The invasion of chick embryonic phragments by malignant cells could be blocked completely by microtubule inhibitors (nocodazole, vinblastine, colchicine) but not by antimetabolites or alkylating agents. These experiments corroborate the high degree of activity of e.g. nocodazole in inhibiting the formation of metastases in the Lewis Lung tumor model. (Atassi and co-workers, 1975). It seems thus logical to conclude that malignant invasion is essentially some type of directional migration (or taxis) which depends on an acto-myosin system for force production and the microtubule system as a coordinating apparatus.

THE ORGANIZED ASSEMBLY OF MICROTUBULES IN LIVING CELLS

The scheme of Fig. 16 represents the turnover of the microtubule system in mammalian cells throughout interphase and mitosis. Two essential features are apparent. On the one hand microtubule based structures are highly organized. On the other hand they are very dynamic and versatile, changing continuously in shape and composition. How are these seemingly paradoxical characteristics married.
From biochemical investigations on isolated tubulin it has become apparent that microtubule assembly can be described in terms of a dynamic equilibrium between oligomers and polymers which is influenced by promoting and inhibiting factors (for review see Dustin, 1978). The organized assembly would depend on the existence of so-called microtubule organizing centres (MTOC) (Pickett-Heaps, 1975) from which microtubules are seen to originate. The best known MTOC's in mammalian cells are the centrosomes consisting of the centriolar duplexes surrounded by an osmiophylic cloud and the kinetochores on the chromosomes.
Experiments with isolated MTOC's and lysed cell models have suggested that these structures fulfil their organizing function by serving as nucleation sites from which microtubules can start to assemble (Burns and co-workers, 1974; Gould and Borisy, 1977; McGill and Brinckley, 1975; Telzer and Rosenbaum, 1975). However, this evidence has recently been submitted to serious criticism by Pickett-Heaps (1978) who proposed that anchoring of preformed microtubules onto the MTOC's could be an equally valid hypothesis. Some attempts have been made to demonstrate nucleated assembly of microtubules in living cells after treatment with colchicine or colcemid using immunofluorescence. The interpretation of these experiments is hampered by technical limitations. (Frankel, 1976; Osborn and Weber, 1976; Spiegelman and co-workers, 1979). Fully active concentrations of colchicine alkaloids are virtually irreversible. Probably for this reason relatively low concen-

trations (10^{-6} - 10^{-7} M) were used. With such concentrations it is highly probable that short microtubule phragments persist in the vicinity of the MTOC's, which escape detection by immunofluorescence. Such microtubule phragments are powerful nucleators themselves and would thus obscure completely the intrinsic activity of the MTOC's. The necessary ultrastructural controls were unfortunately not performed in these reports.

In order to untie this question we have used two new approaches. First nocodazole was used as a fully active but immediately reversible inhibitor of tubulin assembly in cultured mammalian cells. Second the neoformation of microtubules after complete destruction was followed by the PAP-immunocytochemical procedure (De Brabander and co-workers, 1977, 1979) combined with toluidine blue staining of the chromosomes and ultrastructural investigation.

The experimental procedure can shortly be summarized as follows. Mouse embryonal 3T3-type cells (De Brabander and co-workers, 1976) and rat kangoroo (PTK2) cells were cultured in Eagle's minimal essential medium supplemented with 10 % fetal bovine serum as described previously (De Brabander and co-workers, 1976, 1977).

Coverslip cultures were fixed and processed for immunocytochemical demonstration of microtubules exactly as described previously (De Brabander and co-workers, 1977). In short: fixation was done with 0.3 % glutaraldehyde for 15 min at room temperature. This was followed by dehydration and rehydration in acetone (50 %, 100 %, 50 % each for 5 min).
Then the preparations were incubated with an affinity purified antibody against rat brain tubulin (De Brabander and co-workers, 1977), followed by the bridge antibody (goat anti rabbit IgG) and the PAP-complex (Sternberger, 1974). The reaction with 3,3-diaminobenzidine was performed according to Graham and Karnovsky (1966) and followed by osmification. Hereafter, the cells were counterstained with toluidine blue as described in order to contrast the chromosomes (De Brabander and co-workers, 1979). The coverslips were then mounted and viewed with an ordinary light microscope.

For electron microscopy cultures were processed by conventional techniques exactly as described previously. (De Brabander and co-workers, 1976). Thin sections were viewed in a Philips EM 300 or EM 201.

The following experiments were conducted: MO-cells and PTK2 cells were treated with nocodazole (10, 1 and 0.1 μg/ml) for 4 h and 24 h. The medium was decanted, the cultures were washed 3 times with 15 min intervals, with complete growth medium at 36°C, and further incubated. Control and treated cultures were fixed before and 15, 30, 60, 120 min 4 h and 24 h after the first wash.
Parallel cultures were processed for light microscopic immunocytochemistry, combined with toluidine blue staining and for electron microscopy. The data thus obtained on large populations of fixed cells were correlated to time lapse microcinematographic observations of individual mitoses treated at various stages (pro/meta/ana/telo) with different concentrations of nocodazole with and without subsequent removal of the compound.

Microtubule assembly in interphase cells

Interphase MO-cells contain a very dense radiating complex of cytoplasmic microtubules (Fig. 1) which is highly sensitive to nocodazole (Fig. 2). Treatment with 10 or 1 μg/ml induces a complete destruction also in the vicinity of the centrosomes. Virtually immediately after the compound has been washed out microtubules are repolymerized forming a completely disorganized mesh which is difficult to visualize in the immunocytochemical preparations (Fig. 3). Ultrastructural observations confirm the presence of many microtubule phragments throughout the cytoplasm (Fig. 3).

Besides the mesh of randomly polymerized microtubules small asters appear around the centrosomes, which are often located in a peripheral fashion. Within time the microtubule system is reorganized into a radiating complex emanating from the centrosomes which have migrated towards the nuclear vicinity. The first cells with an organized system appear after 4 h only. After 24 h most cells have assumed a completely normal cytoplasmic microtubule complex. It cannot be concluded whether the reorganization is due to the rearrangement of the random microtubules or to the continued centrosomal nucleation accompanied by disintegration of the random microtubules.

Interphase PTK2 cells have a delicate interphase microtubule network (Fig. 5). This is however much more resistant to nocodazole than in MO-cells. Even with 10 μg/ml some cells still contain a small dot which is shown at the ultrastructural level to consist of the centrosome surrounded by short microtubules. Reversal of nocodazole treatment results in a pronounced centrosome associated microtubule assembly (Fig. 7). This allows the extremely fast (within 15 - 60 min) reconstruction of a complete organized system (Fig. 8). Careful observation however shows that random assembly, not associated with the centrosomes does occur in a considerable number of PTK2 cells also (Fig. 8).

Fig. 1. MO-cell stained with the PAP-antitubulin procedure, showing an extensive cytoplasmic microtubule complex (x 350).

Fig. 2. MO-cell treated with nocodazole (1 μg/ml) for 24 h, and stained with the PAP-antitubulin procedure. Note the complete absence of microtubules (x 450).

Fig. 3. MO-cell treated with nocodazole (1 μg/ml; 24 h) 30 min after reversion, stained with the PAP-antitubulin method. Note the appearance of an orderless network (x 350).

Fig. 4. Electron micrograph of an MO-cell treated with nocodazole (1 μg/ml; 24 h) 30 min after reversion. Note the presence of numerous microtubules in the peripheral cytoplasm (x 15.000).

Fig. 5. PTK2 cell stained with the PAP-antitubulin procedure, showing a delicate but highly organized cytoplasmic microtubule complex (x 850).

Fig. 6. PTK2 cells treated with nocodazole (10 μg/ml; 4 h) and stained with the PAP-antitubulin procedure. Note the complete absence of microtubules (x 700).

Fig. 7. PTK2 cells treated with nocodazole (10 μg/ml; 24 h) 60 min after reversion, stained with the PAP-antitubulin procedure. Note the appearance of 1 or 2 asters radiating from focal points (x 900).

Fig. 8. PTK2 cell treated with nocodazole (10 μg/ml; 4 h) 30 min after reversion, stained with the PAP-antitubulin procedure. Besides the strong centrosomal nucleation numerous short microtubule phragments are seen in the peripheral cytoplasm (x 1.000).

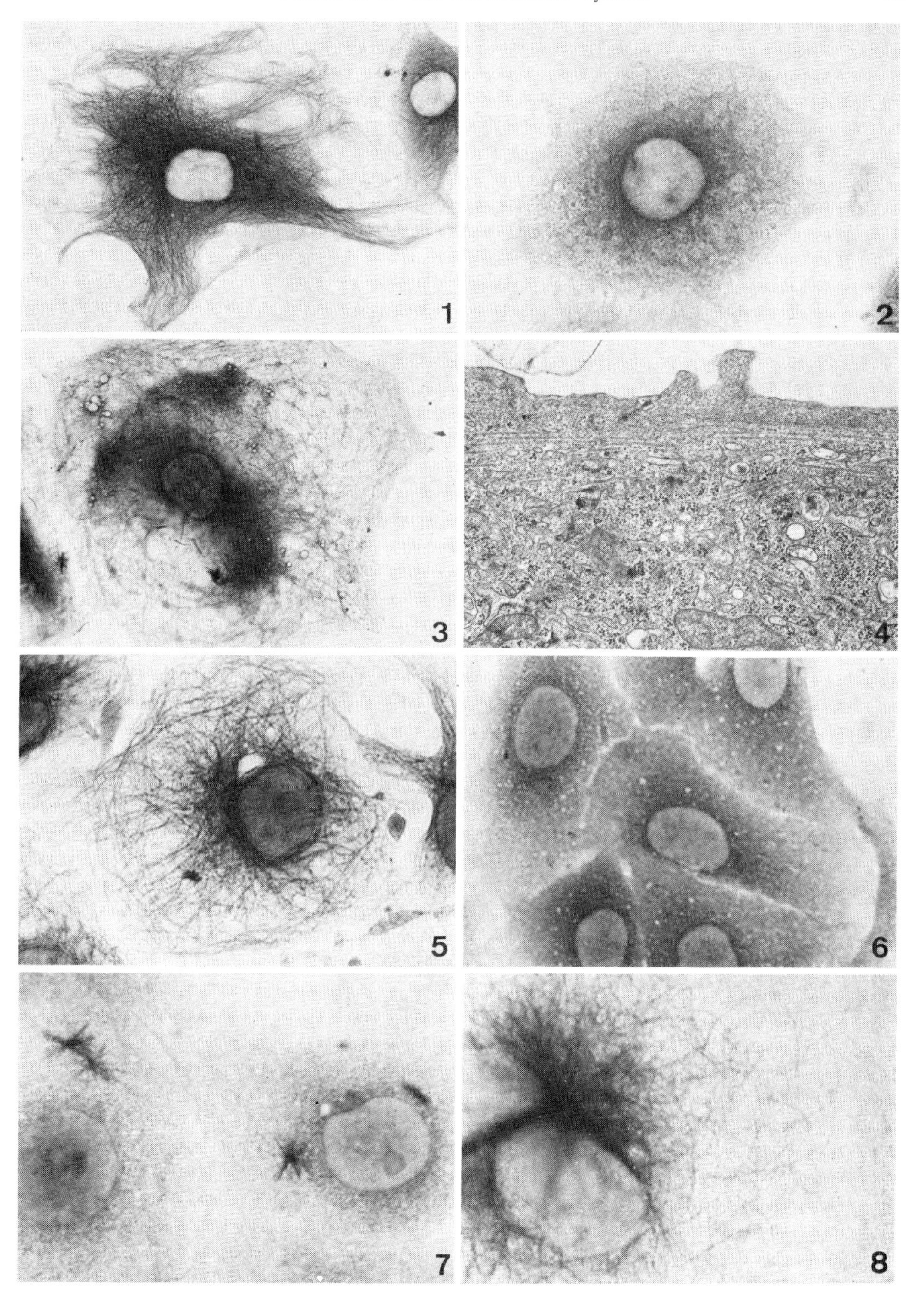
1
2
3
4
5
6
7
8

From these data the following conclusions can be drawn. In the present experimental situation both random assembly and centrosome associated assembly can occur. Other nucleation sites could not be detected in interphase cells. The prependorance of either nucleated or random assembly depends strongly on the cell type. This may be due to differences in tubulin concentration, the balance between inhibiting and promoting factors and the nucleating "strength" of the centrosome. The prevalence of centrosomal assembly strongly favors the rapid reconstruction of an organized cytoplasmic microtubule system. It is thus highly probable that in untreated cells, which show an organized microtubule pattern, nucleation is largely limited to the centrosomal region. The random assembly in nocodazole treated cells after reversion is most probably due to the sudden release of concentration of polymerizable tubulin which exceeds the threshold above which spontaneous nucleation can occur, a situation which apparently does not occur at any point during the normal cell cycle.

Microtubule assembly in mitotic cells

Mitotic MO-cells and PTK2 cells are both equally sensitive to the action of nocodazole. With all three concentrations the spindle is completely destroyed and the chromosomes are distributed in an orderless fashion (Fig. 9). Ultrastructural observation shows that a concentration of 1 µg/ml suffices to induce absence of microtubules even in the vicinity of the centrosomes and the kinetochores (Figs. 10, 11). Release of the nocodazole block results in nucleation of microtubules at the kinetochores and centrioles within 15 min. The localized nucleation is easily visible in the light microscope (Figs. 12, 13). Ultrastructural observation of semi-serial sections confirms that one can assume with a high degree of certainty that nucleation is limited to these structures (Figs. 14, 15). Indeed, microtubules are found exclusively in their vicinity and not one is seen in the peripheral cytoplasm.

The ultrastructural pictures disclose that the short microtubule phragments which are formed initially do not form an organized array pointing away from the kinetochores or centrosomes. Instead, an intricate and apparently unorganized mesh of tubules is seen (Fig. 14). This suggests to us that the nucleation is not a strictly local phenomenon but that the MTOC's are surrounded by an ill defined region in which tubulin assembly is favored.

Fig. 9. Electron micrograph of a mitotic PTK2 cell treated with nocodazole (1 µg/ml; 4 h). Note the orderless distribution of the chromosomes and the absence of spindle microtubules (x 7.500).

Fig. 10. Electron micrograph of a mitotic PTK2 cell treated with nocodazole (1 µg/ml; 24 h). The two kinetochores (arrowheads) of the chromosome are completely free of microtubules (x 19.000).

Fig. 11. Electron micrograph of a mitotic PTK2 cell treated with nocodazole (0.1 µg/ml; 24 h). The centriole (arrowhead) and the pericentriolar material are devoid of microtubules (x 27.500).

Fig. 12-13. Counterdrawings made from colour micrographs of mitotic PTK2 cells stained with the PAP-antitubulin procedure and toluidine blue. Fig. 12 shows a cell 15 min after release from nocodazole block (1 µg/ml; 24 h). Each chromosome shows a darkly staining tuft of microtubules at the centromere region. Isolated small asters (arrowheads) correspond to the displaced centrosomes.
Fig. 13 shows a cell 30 min after release from nocodazole block (1 µg/ml; 24 h). The newly formed microtubule bundles have elongated preferentially between the centrosomes (arrowheads) and the chromosomes.

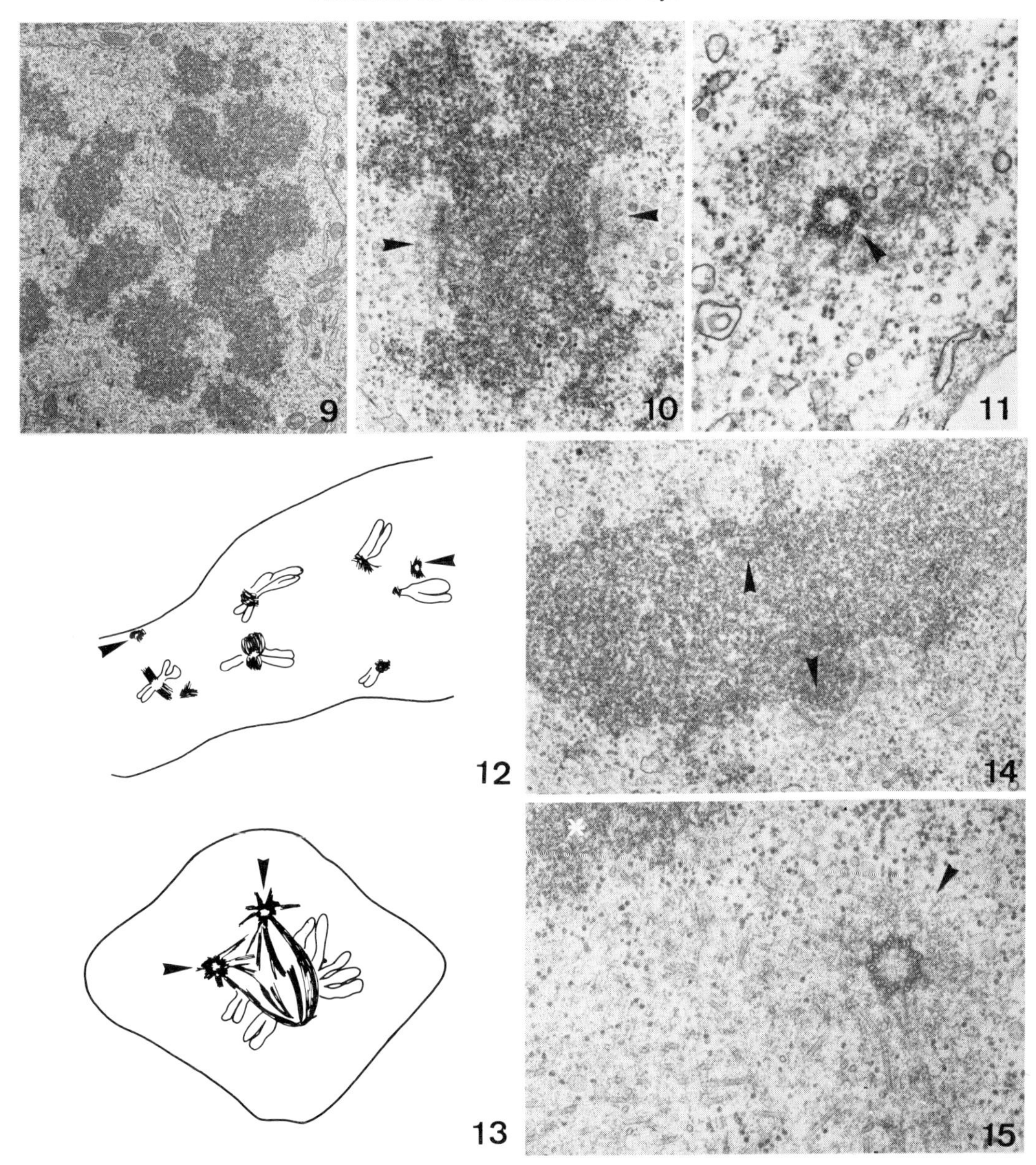

Fig. 14. Electron micrograph of a mitotic PTK2 cell 15 min after release from nocodazole block (1 μg/ml; 4 h). Note the appearance of numerous microtubules exclusively in the vicinity of the kinetochores (arrowheads) (x 25.500).

Fig. 15. Electron micrograph of a mitotic PTK2 cell 4 h after release from nocodazole block (10 μg/ml; 4 h). Numerous microtubules appear around the centriole (arrowhead). Most of them are oriented preferentially towards the chromosomes (asterisk) which are located at the left side of the picture (x 30.500).

With the three concentrations and 2 times of treatment nucleation is seen in 100 % of the cells whose chromosomes are in the metaphase configuration. In prophase cells, centrosomal nucleation but no kinetochore nucleation is seen. This confirms our previous observations on normal mitoses (De Brabander and co-workers, 1979). During the further observation period the microtubule bundles elongate and apparently interact with each other giving rise to the formation of complex patterns (Fig. 13). This evolves finally in the reconstruction of normal metaphase, spindles which evolve then through anaphase and telophase. Counts of normal and abnormal mitoses at different time intervals after reversion showed that e.g. in the 1 μg/ml - 4 h group most normal mitoses that appear after 1 - 2 h are derived from previously blocked mitoses and not from new mitoses that started after the reversion.
The resumption of normal mitosis after nocodazole block could also be established positively by time lapse microcinematography.

Although the released mitotic cells start thus from very complex and variable situations, with regard to the intracellular localization of the MTOC's, they are apparently mostly able to gain the normal metaphase configuration. Metaphase can thus be regarded as an equilibrium situation which is reached inevitably, to a large extent independently from the previous history of the cell. It is difficult to conceive possible mechanisms that could explain this inherent organizational power solely on the basis of non-oriented microtubule nucleation by the MTOC's. Careful observation of the initial stages in the reconstruction of the spindle might give some additional clues. Indeed, the initial nucleation of small microtubule phragments is apparently isometric, occurring to the same degree all around the centrosomes and kinetochores. The further elongation of microtubule bundles however shows a different pattern. Apparently the bundles elongate preferentially between neighbouring MTOC's regardless of how complex the interrelationships may be (Fig. 13). Pronounced centrosome to centrosome and centrosome to kinetochore bundles appear thus very quickly while the microtubules pointing to other directions do apparently not elongate.
We interpret this preferential assembly of inter-MTOC bundles as reflecting a summation of the assembly promoting capacity of neighbouring MTOC's.

DISCUSSION

We believe that these experiments provide the first sufficiently controlled data on the reassembly of the microtubule system in mammalian cultured cells following its complete destruction by a specific and reversible inhibitor of tubulin assembly. The essential technical factors in our approach which differ from previous experiments (Osborn and Weber, 1976; Frankel and co-workers, 1976; Brinckley and co-workers, 1976; Spiegelman and co-workers, 1979) are the following. The use of nocodazole as a specific (unlike cold) but immediately reversible (unlike colchicine) inhibitor. The fine structural control of complete microtubule disassembly. The fixation of the cells with a procedure that preserves the fine structure of microtubules intact. The observation of reassembly by a combined approach using a stable immunocytochemical labelling procedure, coupled with toluidine blue staining of the chromosomes and ultrastructural confirmation of the light microscopic data.

The most salient points can be summarized as follows. In interphase cells released from nocodazole block, spontaneous non centrosomal, assembly can occur. This is most probably accompanied by centrosomal nucleation. The balance between spontaneous assembly and centrosomal assembly is dependent on the cell type and experimental variables. Preponderance of centrosomal nucleation is associated with a rapid reconstruction of an organized cytoplasmic microtubule complex.

In the mitotic cells, released from nocodazole block, microtubule nucleation is strictly limited to the vicinity of the kinetochores and centrosomes, although they

were completely free of microtubule remnants. We believe that this shows unequivocally that these MTOC's can indeed nucleate tubulin assembly in living cells. It is thus highly probable that this occurs also in normal mitosis.
This conclusion is valid only if one defines nucleation in a broad sense as: the capacity to induce tubulin assembly in a more or less localized region within a general environment where assembly does not occur. The data do not provide evidence that the kinetochores or the centrosomes actually contain seeds or templates onto which tubulin is assembled.

Most discussions concerning the molecular mechanisms through which kinetochores and centrosomes nucleate the assembly of microtubules assume indeed rather tacitly that it is a strictly local phenomena. In analogy to the situation encountered in the formation of cilia and flagella onto basal bodies (Rosenbaum and co-workers, 1975), they would act as seeds enabling the formation of short microtubule ends that elongate by addition of dimers at the distal ends (Borisy, 1978). While this may be one aspect of the MTOC function it is unable to explain an important event which is the preferential elongation of microtubules along the axis between two nucleation sites, e.g. between 2 centrosomes or a centrosome and a kinetochore. This is a constant observation as well in normal mitosis as in mitotic reorganization after nocodazole block and it is probably essential in the organized contruction of the spindle. The organizing role of the MTOC's could be much more refined if one would assume that they function, at least in part, by creating a surrounding gradient in which the threshold for tubulin assembly is lowered with respect to the general, inhibitory cytoplasmic environment. Then one would expect microtubule elongation to be favored between two MTOC's where neighbouring gradients overlap. The existence of factors in the cytoplasm that inhibit spontaneous assembly has already been evidenced by Bryan and co-workers (1975).
The presence of an even higher concentration of such factors in mitotic cells is substantiated to some extent by our earlier observations that vinblastine induced tubulin paracrystals disappear during mitosis and reappear afterwards (De Brabander and co-workers, 1979). Moreover, it is also suggested by the experiments described here which show that, after release from nocodazole block, random assembly does occur in interphase cells but never in mitotic cells.

The existence of a threshold-lowering gradient around MTOC's is purely hypothetical. It would however fit in, much better than a strictly local template function, with a number of facts such as: - the existence of free microtubules, not anchored to any MTOC (McIntosh and co-workers, 1975) - the morphological appearance of newly nucleated microtubules after nocodazole block which are located in the vicinity of the nucleation sites but usually not really attached onto them. - the higher resistence of microtubules against exogenous inhibitors (nocodazole, colchicine) in the vicinity of the MTOC's and in particular between neighbouring MTOC's.

The gradient concept does not necessarily exclude a concomitant anchoring role of the MTOC's. (Pickett-Heaps and Tippit, 1978). In fact a most plausible machinery could be imagined in this way: the peri-MTOC polymerization promoting gradient produces many short microtubules which are then anchored into the MTOC itself and able to elongate further mainly at the non-anchored end. The gradient concept, however, seems sufficient but essential to explain in interplay with fluctuating cytoplasmic concentrations of inhibitory factors in a logical, smoothly running machinery, all the transitions of microtubule based structures of mammalian cells throughout the cell cycle. The following sequence of events may be envisaged to occur (Fig. 16). The generally radiating aspect of the interphase microtubule system may very well reflect a dynamic equilibrium between inhibitory factors and the promoting gradient emanating from the centrosome. In prophase the cytoplasmic microtubule complex shortens, maybe by the release of inhibiting factors from the nucleus while its membrane is opened. This could correlate with the rather general absence of microtubules from interphase nuclei in higher eukaryotes. In any case, a relatively high concentration of inhibi-

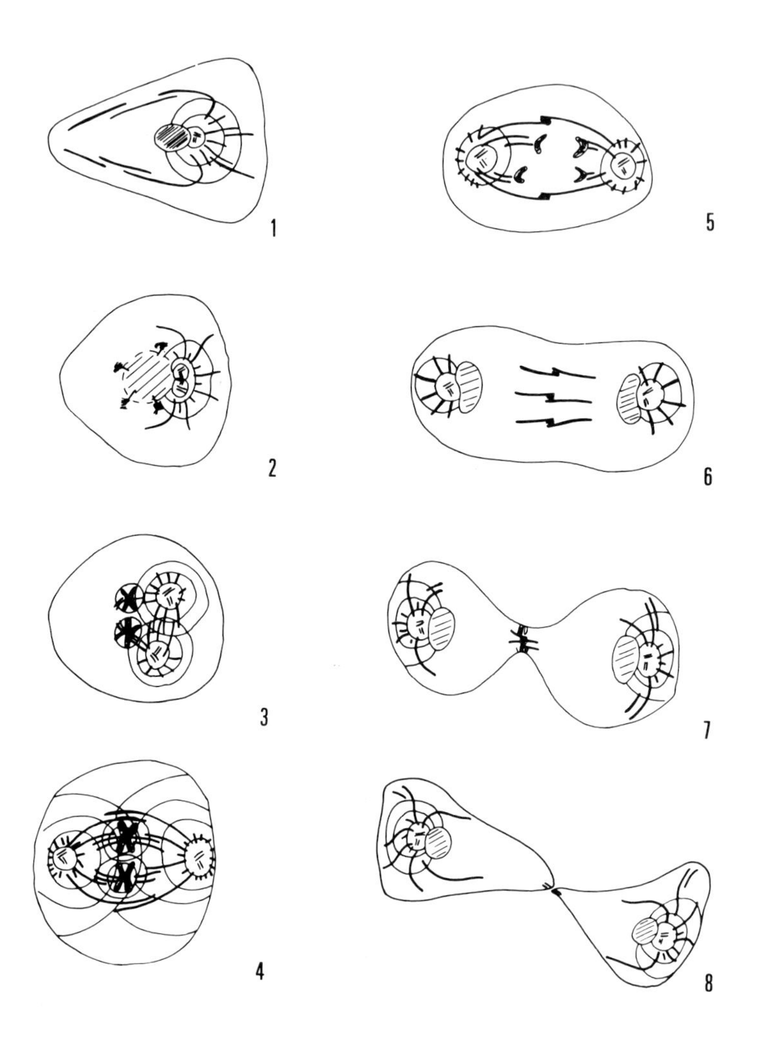

Fig. 16. Schematic representation of the evolution of the microtubule system throughout the cell cycle. Microtubules are presented by the heavy lines. The hypothetical polymerization promoting gradient around the centrosomes (// =) and the chromosomes ( ) is represented by the finely drawn circles. In interphase (1) microtubule nucleation is largely limited to the centrosomal region. In prophase (2) the CMTC shortens and nucleation is favored between the reduplicated centrosomes. During prometaphase (3) further elongation is favored between the centrosomes, which separate, and between the centrosomes and the kinetochores which have ripened. The system reaches equilibrium at metaphase (4). During ana-telophase (5-7) the interdigitating sets of microtubules slide past each other resulting in spindle elongation and chromatid separation. After telophase (8) the CMTC is reconstructed radiating from the centrosomes.

ting factors during mitosis would strongly suppress random, non-nucleated, tubulin assembly which is a prerequisite for the ordered construction of the spindle. The result is that the cell is left with a relatively small aster of microtubules around the now reduplicated centrosomes. Polymerization is now strongly favored between the two centrosomes where the two promoting gradients overlap. This is aided by a higher tubulin concentration, derived from the peripheral microtubules. Continuing polymerization in a dynamic equilibrium fashion results thus in the separation of the two asters. The tendency of the two sets of interdigitating microtubules to associate laterally probably aids in the formation of the premetaphase spindle.
The inclusion of the chromosomes in the spindle would be ascertained by at least two factors. First, relative cell rounding, which is a rather general event in cells with open spindles, ensures adequate contact. Second, during prometaphase the kinetochores ripen and start to nucleate. Through the influence of overlapping gradients the association with centrosome derived bundles is favored. Continuing polymerization of the kinetochore and centrosome derived microtubule sets until equilibrium is reached would inevitably result in the adoption of the metaphase morphology with the chromosomes located at the equatorial region. The construction of the metaphase spindle would thus depend largely on dynamic equilibrium reactions (Inoue and Sato, 1967) and lateral association of microtubule bundles. This is confirmed by time lapse observations showing the extreme sensitivity of this process to inhibitors of tubulin polymerization such as nocodazole, colchicine etc. On the contrary, the structural and functional integrity of the metaphase spindle and later stages are highly resistant to nocodazole and other inhibitors (Oppenheim and co-workers, 1973). For instance, 10 μg/ml nocodazole does not block post-metaphase movements while 0.1 μg/ml suffices to block completely pre-metaphase stages.
This implies that from metaphase on spindle microtubules have more or less stabilized. Spindle elongation and chromosome separation would thus largely be completed by some kind of sliding of interdigitating microtubule sets, with or without the aid of other molecules (actin, dynein) or by other types of interactions (zipping?) (for review see Pickett-Heaps and Bajer, 1978).

Telophase is characterized by the envelopment of the intermingled chromatids into a new nuclear membrane. This leaves the centrosomes as the only nucleation sites. And indeed, the aster microtubules, emanating from them start to elongate in all directions and reconstruct a radiating cytoplasmic microtubule complex. Elongation of these microtubules far away from the centre of the promoting gradient is probably aided by the raised tubulin concentration (derived from the disintegrating kinetochore fibers) and by a presumptive lowering of the concentration of inhibiting factors to the interphase level. The latter may be achieved simply by assuming a limited biological half life of the factors released during prophase.

In conclusion, the hypothetical sequence of events depicted above, is to our knowledge not contradicted by experimental evidence as far as mammalian cells are concerned. On the contrary, it combines several existing theories on the mechanisms of mitosis for each of which firm arguments have been gathered (see Pickett-Heaps and Bajer, 1978, for review). The basic factors which allow one to propose this unifying concept are the positive evidence for the nucleated assembly in living cells and the suggestive evidence for the preferential elongation of inter-MTOC microtubules which leads to the assumption that the nucleation sites do not merely function as local seeds but create a gradient in which assembly is favored, against a general inhibiting environment.

REFERENCES

Atassi, G., C. Schaws, and H.J. Tagners (1975). *Eur. J. Cancer, 11*, 609-615.
Borisy, G.G. (1978). *J. Mol. Biol., 124*, 565-570.
Bryan, J., B.W. Nagle, and K.H. Doenges (1975). *Proc. Natl. Acad. Sci. USA, 72*,

3570-3574.
Burns, R.G., and D. Starling (1974). *J. Cell Sci.*, *14*, 411-420.
De Brabander, M.J., R.M.L. Van de Veire, F.E.M. Aerts, M. Borgers, and P.A.J. Janssen (1976). *Cancer Res.*, *36*, 905-916.
De Brabander, M., J. De Mey, R. Van de Veire, F. Aerts, and G. Geuens (1977). *Cell Biol. Int. Reports*, *1*, 453-461.
De Brabander, M., J. De Mey, M. Joniau, and S. Geuens (1977). *J. Cell Sci.*, *28*, 283-301.
De Brabander, M., G. Geuens, J. De Mey, and M. Joniau (1979). *Biol. Cell*, *34*, 83-96.
Dustin, P. (1978). *Microtubules*. Springer-Verlag Berlin, Heidelberg, New York.
Frankel, F.R. (1976). *Proc. Natl. Acad. Sci. USA*, *73*, 2798-2802.
Gould, R.R., and G.G. Borisy (1978). *Exp. Cell Res.*, *113*, 369-374.
Graham, R.C., and M.J. Karnovsky (1966). *J. Histochem. Cytochem.*, *14*, 291-304.
Inoue, S., and H. Sato (1967). *J. gen. Physiol.*, *50*, 259-292.
Lazarides, E., and J.P. Revel (1979). *Sci. Amer.*, *240*, 88-100.
Mareel, M.M.K., and M.J. De Brabander (1978). *J. Natl. Cancer Inst.*, *61*, 787-792.
McGill, M., and B.R. Brinkley (1975). *J. Cell Biol.*, *67*, 189-199.
McIntosh, J.R., W.Z. Cande, and J.D. Snyder (1975). Structure and physiology of the mammalian mitotic spindle. In S. Inoue and R.E. Stephens (Eds.), *Molecules and Cell Movement*, Raven Press, New York pp. 31-74.
Oppenheim, D.S., B.T. Hauschka, and J.R. McIntosh (1973). *Exp. Cell Res.*, *79*, 95-105.
Osborn, M., and K. Weber (1976). *Proc. Natl. Acad. Sci. USA*, *73*, 867-871.
Pickett-Heaps, J.D. (1975). *Ann. N.Y. Acad. Sci.*, *253*, 352-361.
Pickett-Heaps, J.D., and A.S. Bajer (1978). *Cytobios*, *19*, 171-180.
Pickett-Heaps, J.D., and D.H. Tippit (1978). *Cell*, *14*, 455-467.
Rosenbaum, J.L., L.I. Binder, S. Granett, W.L. Dentler, W. Small, R. Sloboda, and L. Haimo (1975). *Ann. N.Y. Acad. Sci.*, *253*, 147-177.
Spiegelman, B.M., M.A. Lopata, and M.W. Kirschner (1979). *Cell*, *16*, 239-252.
Sternberger, L.A. (1974). *Immunocytochemistry*. Englewood, New Jersey: Prentice-Hall.
Telzer, B.R., M.J. Moses, and J.L. Rosenbaum (1975). *Proc. Natl. Acad. Sci. USA*, *72*, 4023-4027.

Cell Movement *in vitro*

Essay of Characterization of Skin Fibroblasts derived from Patients with Mammary Tumours. II - Biological Properties of Cell Cultures from Different Body Sites

B. Azzarone, M. Mareel* and A. Macieira-Coelho

I.C.I.G. (INSERM U50), 94800 Villejuif, France
**Clinic for Radiotherapy and Nuclear Medicine, Ghent, Belgium*

ABSTRACT

We compared the growth properties of two skin fibroblastic cultures obtained from different body sites of a patient with breast cancer. Both cell lines are capable to grow with the same efficiency in soft agar and in presence of low serum concentration. In both cell lines the maximal saturation density increases when serially transferred at a 1:1 split ratio. However the analysis of growth and survival curves and of DNA synthesis reveals the existence of different proliferative capacities. We also analyzed the kinetics of entrance into S period of cultures maintained at different split ratios, but plated at the same initial inocula at the time of the experiment. Our data suggest that subcultivation at 1:1 split ratios recruits short cycling cells with greater growth potentials.

KEY WORDS

Cancer patient; tissue culture; skin fibroblasts; transformation; growth kinetics.

INTRODUCTION

We have previously compared the growth properties of mammary skin fibroblasts, derived from patients with breast cancer, with those of fibroblastic cultures, derived from patients with mammary breast lesions.

The parameters analyzed included : percentage of cells synthesizing DNA as a function of cell density, increase of maximal cell density by repeated transfers at a 1:1 split ratio, growth in semisolid medium and growth in low serum concentration (Azzarone and Macieira-Coelho, 1979).

We have found that fibroblastic cultures from cancer patients respond in an abnormal way to each biological parameter tested, while mammary skin fibroblasts from patients with benign lesions show some but not all of these abnormal growth properties.

This altered behavior of fibroblasts in vitro is now approached further comparing the kinetics of proliferation of skin fibroblastic cultures obtained from different body sites of the same patient with a breast cancer and maintained at different cell densities.

MATERIALS AND METHODS

Cell Culture

Skin specimens, derived from the mammary gland (CM19) and from the abdomen (CMV) of a 36 year-old patient with breast cancer, were explanted for cell culture according to a previously described technique (Azzarone and others, 1976b). Cells were maintained in Eagle's minimum essential medium supplemented with 10% fetal calf serum (F.C.S.) and 16 μCi gentamycin/ml. The cultures were carried in 75 cm^2 plastic bottles. When the cells formed a confluent sheet and no mitoses could be observed, they were subcultivated into new bottles at a 1:2 split ratio. The nutrient medium was left unchanged between each subcultivation. Cell counts were performed after trypsinization with an electronic Coulter counter.

Comparison between Growth Fraction and Cell Density

Cells were suspended at different concentrations in 2.5 ml of culture medium and seeded into 30 mm plastic Petri dishes. Special care was taken to ensure equal and uniform cell density at the start of an experiment, in all culture dishes belonging to the same group. Therefore cell counts were performed on cell suspension immediately before and after seeding. Each day after seeding, the cells of two cultures were trypsinized and counted until no further increase in cell number was observed without medium change. The counts in duplicate dishes differed by less than 10% throughout the whole experiments. ^{3}H-TdR in a final concentration of 0.1 μCi/ml (s.a. 2 μCi/mmole) was added to duplicate sister cultures. Each day after subcultivation and 24 h later the cell cultures were prepared for autoradiography, according to a previously described technique (Macieira-Coelho and others, 1966).

Unstability of Post-confluent Saturation Density

Confluent cultures were trypsinized, counted and divided into two groups; in one each culture was subcultivated into two new bottles (1:2 split ratio) and in the other one each culture was subcultivated into one new bottle (1:1 split ratio).

Anchorage Dependence

Colony formation in soft agar was performed according to a previously described technique (McAllister and others, 1967).

Growth in Low Serum Concentration

Confluent cultures were trypsinized and the cells were seeded at a concentration of $1x10^6$ cells/30 ml in 75 cm^2 culture flasks in the presence of 1% F.C.S. After a 24 h-period, cultures were rinsed and divided into two groups, one was refed with Eagle's MEM supplemented with 10% F.C.S. and the other one was refed with 2% F.C.S. When the cultures supplemented with 10% F.C.S. became confluent and no mitoses could be seen, the two groups were trypsinized and counted.

RESULTS

Survival curves (fig. 1) of the skin fibroblastic cultures CM19 and CMV are different although the cells are morphologically similar. Indeed CM19 cultures reach higher saturation densities. In both cell lines there is an increase of the maximal cell density when they are serially subcultured at a 1:1 split ratio, however

CM19 cells achieve saturation density levels higher than those of CMV cells.

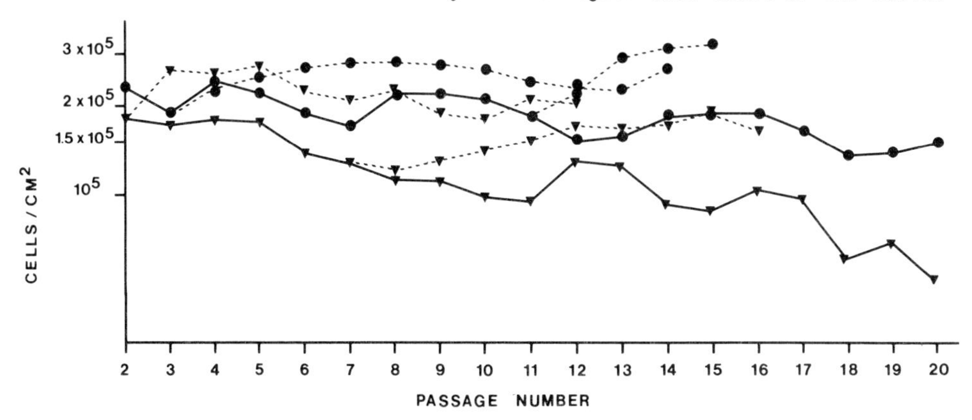

Fig. 1. Survival curves of CM19 cells transferred at a 1:2 (●—●) or 1:1 (●---●) split ratios and of CMV cells transferred at 1:2 (▼—▼) or 1:1 (▼---▼) split ratios.

Figures 2 and 3 illustrate the growth and DNA synthesis curves of CMV and CM19 cell cultures at the 5th passage. Cells from the same culture were seeded at different densities in order to study the influence of the inoculum size.

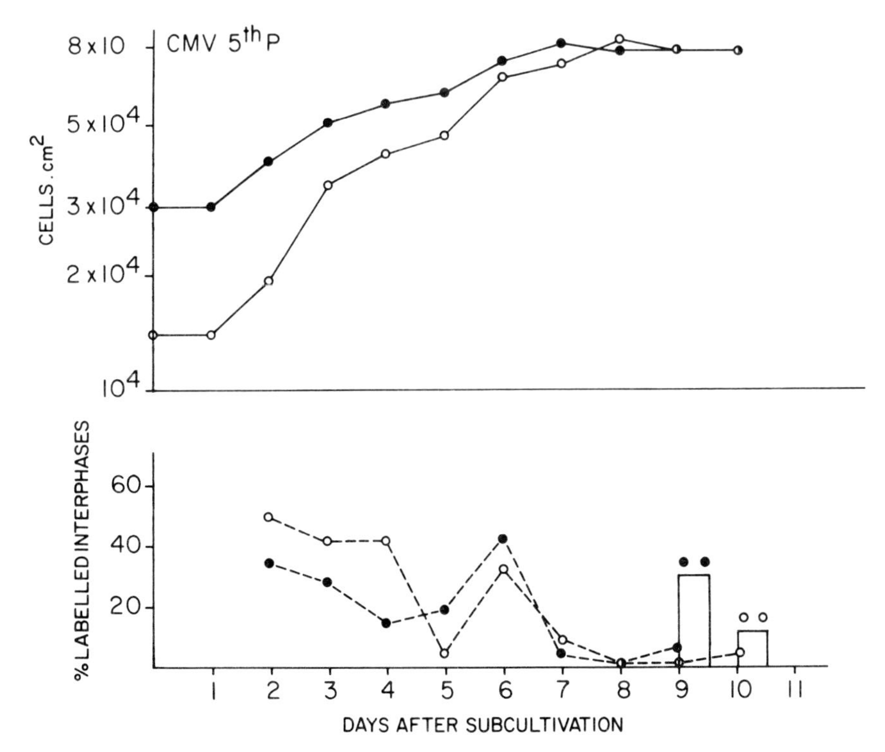

Fig. 2. Cell counts (——) plotted semilogarithmically and percentage of labelled interphases (---) plotted arithmetically on different days after subcultivation. The bars indicate the percentage of labelled interphases along the edge of a wound obtained by scraping with a rubber policeman at the time the cells formed a confluent sheet.

In CMV cells the counts obtained at the time of confluency show that in spite of different inocula the cultures had the same density at the time DNA synthesis approach zero levels. It can also be seen that the maximum number of cells synthesizing DNA during the second day after subcultivation is inversely related to the initial seeding density. This behaviour has been confirmed in similar experiments performed on the 8th and 20th subcultures.

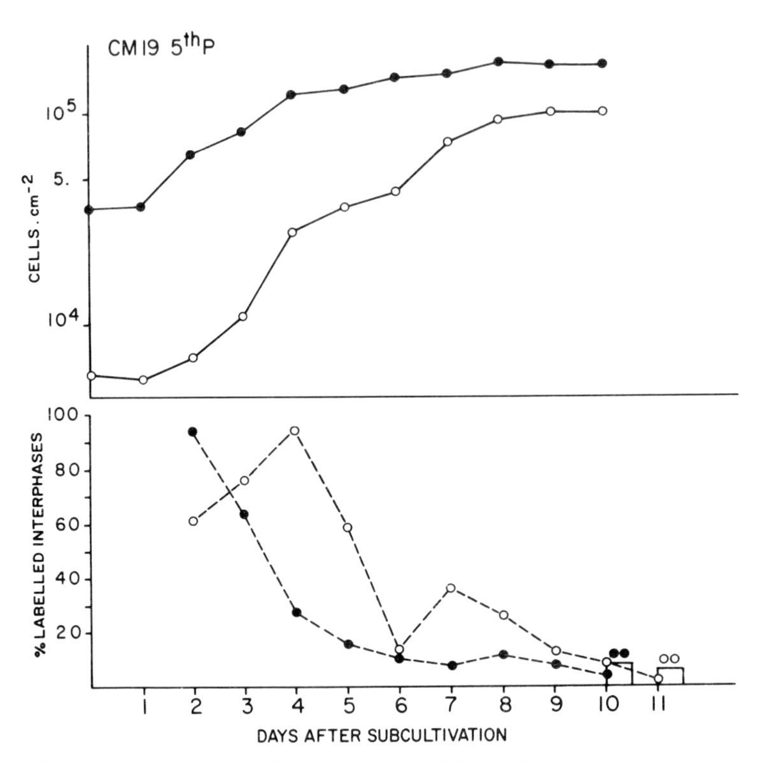

Fig. 3. Cell counts (——) plotted semilogarithmically and percentage of labelled interphases (---) plotted arithmetically on different days after subcultivation. The bars indicate the percentage of labelled interphases along the edge of a wound obtained by scraping with a rubber policeman at the time the cells formed a confluent sheet.

In CM19 cells on the contrary the density at which growth inhibition occurs, declines with a lower inoculum and the percentage of cells synthesizing DNA during the second day after subcultivation increases with a higher inoculum. In these experiments, although the initial inocula were similar in both cell lines, the percentage of labelled interphases during the second day after subcultivation is more elevated in CM19 cultures (92%) than in CMV cultures (40%).

Table 1 shows the time of appearance of the cells entering S phase in CM19 cultures previously propagated at different cell densities and seeded at the same inoculum at the time of the experiment. Resting 19th passage CM19 sister cultures were split 1:1 and 1:2, when cells formed a confluent sheet they were counted ($2x10^5$ cells/cm^2 and $1.4x10^5$ cells/cm^2 respectively) and then seeded at about $3x10^4$ cells /cm^2. Cells were continuously labelled during the first 15 h and then fixed, the remaining cultures were labelled from the 15 h on and fixed at different times thereafter. The percentage of cells synthesizing DNA is low until the 15th h after subcultivation, it is however higher in the plates derived from the 1:1 split cultures. The rate of entrance into the S period is faster in the latter group. It is interesting that the time of appearance and the final percentage of labelled interphases are nearly the same in the 1:2 group (17th passage) and in the 1:1 group (20th passage).

TABLE 1 Time of Appearance of Labelled Interphases at Different Times after Subcultivation in CM19 Cultures Maintained at Different Cell Densities and Seeded at the Same Inoculum

TIME (hours)	PERCENTAGE OF LABELLED INTERPHASES		
	17th passage	20th passage	
	1:2 derived cultures	1:2 derived cultures	1:1 derived cultures
0-15	9.6%	0.6%	6%
19	19.2%	6 %	18%
23	36 %	12 %	33%
27	42 %	22 %	42%
31	53 %	33 %	49%

Results are based on the analysis of 2 000 cells for each sample.

As shown in table 2, both cell lines are able to grow in soft agar and in culture medium supplemented with 2% F.C.S. with similar efficiency.

TABLE 2 Growth in Soft Agar and in 2% F.C.S. of Skin Fibroblastic Cultures Derived from the Mammary Skin (CM19) and from the Abdominal Skin (CMV) of a Patient with a Breast Cancer

Cell line	*Number of colonies in soft agar / 2 x 10^5 cells plated	**Growth response to 2% serum concentration in culture medium after plating 10^6 cells in 75 cm^2 flasks at day zero
CM19	62 (54-71)	2.2 x 10^6 cells/75 cm^2 flask
CMV	62 (58-68)	2.01 x 10^6 cells/75 cm^2 flask

*Each value represents the average of 3 plates. The extreme values batween samples are in parentheses. The number of cells per colony ranges between 10 and 30.

**Each value represents the average of 3 flasks.

Since a good correlation has been found in some cellular models between high saturation densities (Aaronson and Todaro, 1968; Azzarone and Pedullà, 1976a), the ability to grow in semisolid medium (Weinstein and others, 1975) and tumorigenicity in vivo, we decided to test in an in vitro system the invasiveness of our fibroblastic cultures propagated at different cell densities.

We used a method developed by Mareel and coworkers (1979) based on the confrontation of spheroidal aggregates of the cells to be tested, with precultured fragments from embryonic chick tissue in shaker culture. This in vitro assay mimics invasiveness in vitro and a good correlation was found between invasiveness in vitro and formation of invasive tumours in vivo (Mareel, 1979).

Figures 4 and 5 illustrate the different behaviour in a preliminary essay performed on parallel cultures of a cell line (CM11) derived from the skin of a patient with breast cancer that displays different maximal cell densities when subcultured at a 1:2 or 1:1 split ratios.

Fig. 4. Histological section from a 4-day confrontation of a precultured heart fragment with an aggregate of CM11 (1:2 split ratio) cells. CM11 cells have adhered to host tissue without penetrating.

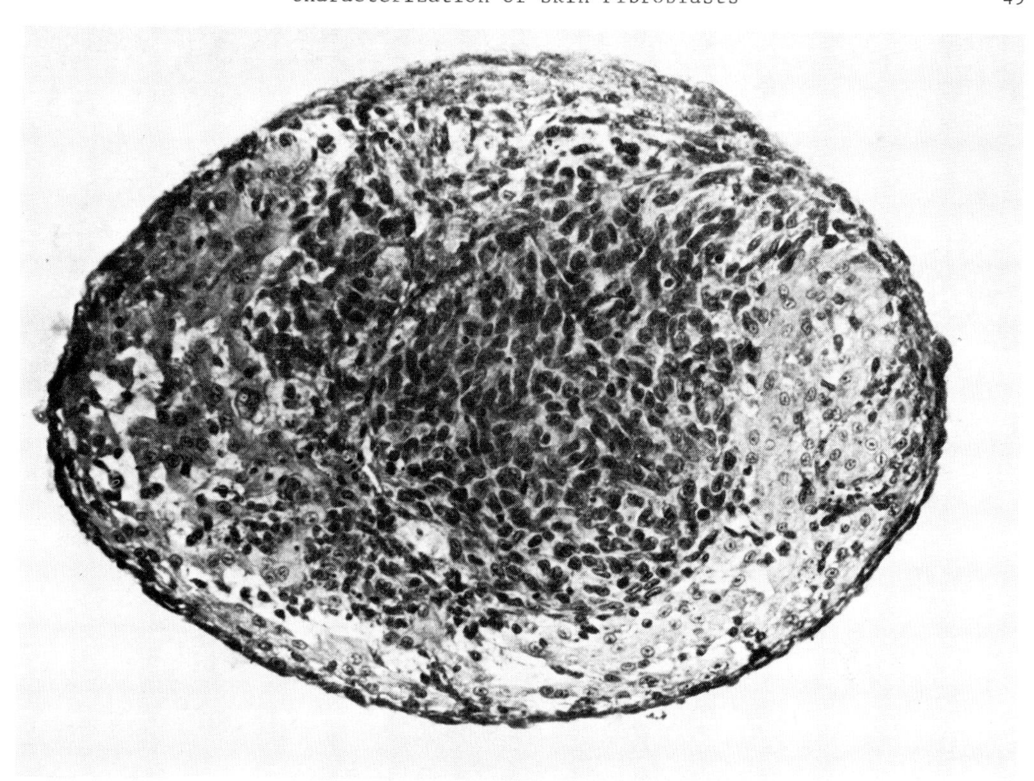

Fig. 5. Histological section from a 4-day confrontation of a precultured heart fragment with an aggregate of CM11 (1:1 split ratio) cells. CM11 cells penetrated the host tissue.

DISCUSSION

It has been reported that skin fibroblasts from patients at high risk of cancer (Miller and Todaro, 1969; Pfeffer and others, 1976) and from cancer patients (Azzarone and others, 1976b; Smith and others, 1976) do not behave in vitro as skin fibroblasts from normal donors. These data suggest that cancer could be also considered as a systemic disease and that skin fibroblasts are a useful tool to study the physiopathology of cancer. However recent data have raised the possibility that human dermis contains several fibroblast-like cells which may differ in behaviour while having a similar morphology (Grove and others, 1977; Kaufman and others, 1975; Harper and Grove, 1979).

In the present paper we have compared the growth properties of mammary skin fibroblasts (CM19) to those of abdominal skin fibroblasts (CMV) derived from the same patient affected with breast cancer. Both cell lines are able to grow with similar efficiency in soft agar and in culture medium supplemented with low serum-concentration. However, comparison of the growth rates at the 5th passage shows significant differences. The fibroblastic cultures from the abdominal skin have growth and DNA synthesis curves that have been previously described as characteristic of normal human fibroblasts (Macieira-Coelho, 1967). Cell division ceases at saturation density that is independent from the inoculum size, and the percentage of cells synthesizing DNA during the second day after subcultivation is inversely related to the seeding density. On the contrary, the fibroblastic cultures from the mammary skin have growth and DNA synthesis curves opposite to those found in CMV cells. Indeed the saturation density at which inhibition occurs is lower after a small inoculum and the percentage of labelled interphases during the second day after subcultivation increases with the inoculum size. The present data is first to show this variation in human skin fibroblasts.

In both cell lines the maximal cell density increases when they are transferred at a 1:1 split ratio, however CM19 cultures reach higher saturation density, confirming their greater proliferative potential. The data recently published by Harper and Grove (1979)could give an explanation for these different growth patterns. These authors have shown that fibroblastic cultures derived from the uppermost region of human dermis (the papillary layer) have greater proliferative capacities when compared to fibroblastic cultures derived from the lowermost region (the reticular layer) of the same skin specimen. Thus it could be argued that CMV cultures have originated from the reticular region of the dermis. This is a reasonable hypothesis which cannot be ruled out. It should be pointed out that the dissection of the skin specimen was done very carefully and only the superficial epidermis region was explanted. During the primary culture an epithelial outgrowth appeared first around each explant, and this suggests that both cell lines originated from the most superficial layer.

In order to study the mechanisms responsible for the different saturation densities, we have compared the time of appearance of cells synthesizing DNA in parallel CM19 cultures seeded at the same inoculum, but derived from cultures which had been previously transferred at 1:1 and 1:2 split ratios. We observed a delayed entrance into the S period and on the second day after subcultivation a lower percentage of cells in S phase in the 1:2-splitted cultures than in the group splitted 1:1. Moreover both the kinetics of appearance and the final number of labelled interphases of the 20th passage, 1:1-derived group are quite similar to those of the 17th-passage cultures derived from a 1:2 split and seeded in the same way (table 1).

Preliminary results with another cell line from another patient with a breast cancer suggest that 1:1 splits select cells capable of penetrating other tissues (fig. 5). Since it has been shown (Macieira-Coelho, 1974 and 1977) that ageing in human fibroblastic cells is accompanied by an increased heterogeneity of cell generation times and to the presence in the late passages of more cells with long cycles, we suggest that subcultivation of CM19 cultures at a 1:1 split ratio favours the proliferation of short cycling cells with higher growth potential.

Although we cannot come to any definitive conclusion concerning the different proliferative capacities of skin fibroblastic cultures from different body sites of the same cancer patient, our data show that these cell lines display some abnormal growth properties commonly used to define transformation, while retaining a normal morphology and a finite lifetime.

ACKNOWLEDGMENTS

This work was supported by contract RCP CNRS 080 571 and by grant CNR 7900601.96, Progetto Finalizzato Controllo Crescita Neoplastica.

REFERENCES

Aaronson, S.A. and G.J. Todaro (1968). Basis for the acquisition of malignant potential by mouse cells cultivated in vitro. Science, 162, 1024-1026.
Azzarone, B. and D. Pedullà (1976a). Spontaneous transformation of human fibroblast cultures derived from bronchial carcinomata. Eur. J. Cancer, 12, 557-561.

Azzarone, B., D. Pedullà and C.A. Romanzi (1976b). Spontaneous transformation of human skin fibroblasts derived from neoplastic patients. Nature, 262, 74-75.
Azzarone, B. and A. Macieira-Coelho (1979). Essay of characterization of skin fibroblasts derived from patients with mammary tumours. Proc. VIIIth Internat. Symp. On the Biological Characterization of Human Tumours, Absctract N°34.
Grove, G., B.A. Houghton, J.W. Cochran, E.D. Kress and V.J. Cristofalo (1977). Hydrocortisone effects on cell proliferation : specificity of response among various cell types. Cell. Biol. Int. Reports,1, 147-151.
Harper, R.A. and G. Grove (1979). Human skin fibroblasts derived from papillary and reticular dermis : differences in growth potential in vitro. Science, 204, 546 -527.
Kaufman, M., L. Pinsky, G. Straisfeld, B. Shanfield and B. Zilahi (1975). Qualitative differences in testosterone metabolism as an indication of cellular heterogeneity in fibroblast monolayers derived from human preputial skin. Exp. Cell Res., 96, 31-36.
Macieira-Coelho, A., J. Pontén and L. Philipson (1966). Inhibition of the division cycle in confluent cultures of human fibroblasts in vitro. Exp. Cell Res., 43, 20-29.
Macieira-Coelho, A. (1967). Influence of cell density on growth inhibition of human fibroblasts in vitro. Proc. Soc. Exp. Biol. Med., 125, 548-552.
Macieira-Coelho, A. (1974) Are non-dividing cells present in ageing cultures? Nature, 248, 421-424.
Macieira-Coelho, A. (1977). Kinetics of the proliferation of human fibroblasts during their lifespan in vitro. Mech. Age. Dev., 6, 341-343.
Mareel, M., T. Kint and C. Meyvisch (1979). Methods of study of the invasion of malignant C3H mouse fibroblasts into embryonic chick heart in vitro. Virchows Arch.B 30, 95-111.
Mareel, M. (1979). Is invasiveness in vitro characteristic of malignant cells? Cell.Biol. Int. Reports, in press.
Mc Allister, R.M., G. Reed and R.J. Huebner (1967). Colonial growth in agar of cells derived from adeno-virus-induced hamster tumours. J. Natl. Cancer Inst., 39, 43-53.
Miller, R.W. and G.J. Todaro (1969). Viral transformation of cells from persons at high risk of cancer. The Lancet, I, 81-82.
Pfeffer, L., M. Lipkin, D. Stutman and L. Kopelovitch (1976). Growth abnormalities of cultured human skin fibroblasts derived from individuals with hereditary adenomatosis of the colon and rectum. J. Cell. Phys., 89, 29-38.
Smith, H.S.,R.B. Owens, A.J. Hiller, W.A. Nelson-Rees and J. Johnston (1976). The biology of human cells in tissue culture. I - Characterization of cells derived from osteogenic sarcoma. Int. J. Cancer, 17, 219-234.
Weinstein, I.B., J.M. Orenstein, R. Gebert, M.E. Kaighn and V.C. Stadler (1975). Growth and structural properties of epithelial cell cultures established from normal rat liver and chemically-induced hepatomas. Cancer Res., 35, 253-263.

A New Approach to the *in vitro* Study of Neoplastic Cell Social Behaviour with a New Light Microscopy Method of Bidirectional Image Transfer

P. Vesely and M. Malý

Institute of Molecular Genetics, Praha 6; Institute of Physics, Praha 8; Czechoslovak Academy of Sciences, Czechoslovakia

ABSTRACT

An approach to the analysis of tumour cell behaviour in vitro relevant to in vivo malignancy is described. It consists of a system of genetically defined inbred and congenic rat strains in which the criterion of malignancy is the propensity of tumour cells to grow across histocompatibility barriers of different strength (Křen and colleagues, 1979), and of a new light microscopy method of bidirectional image transfer, which enables simultaneous observation of tumour cell behaviour at two different magnifications and/or in two focal levels. This opens new possibilities for cinemicrographic studies of the dynamics of cell behaviour in vitro. Three sequences of paired pictures of cells obtained with the method are presented.

KEYWORDS

Tumour cell social behaviour in vitro; assessment of tumour growth capacity in vivo; bidirectional image transfer in light microscopy.

INTRODUCTION

The question whether there is a significant relationship between the type of locomotion of tumour cells with its specific regulation in vitro and invasiveness of the same malignant cells in vivo is still unanswered. Abercrombie and Ambrose (1962) considered the invasion to be an active process involving locomotion. These notions stemmed from earlier results of Abercrombie's group which defined the phenomenon of contact inhibition of locomotion (CIL) of normal fibroblast in vitro and its lack in heterotypic collisions between malignant and normal cells. Abercrombie and his colleagues devoted their further efforts to the detailed elucidation of the mechanism of CIL in normal cells (Abercrombie and Dunn, 1975) and provided substantial evidence on the various qualities of the course of events when tumour and normal cells collide in vitro, such as nonreciprocal contact inhibiton described by Heaysman (1970). They also confirmed the increased motility and randomness of migration of tumour cells in vitro (Abercrombie and Turner, 1978). This correlates well with the data of Wood and co-workers (1967) on the behaviour of V 2 rabbit carcinoma cells filmed for long periods in rabbit ear chamber. Ambrose's group directed their interest to a three-dimensional in vitro models for the study of the mechanisms involved in invasion. Their efforts finally yielded evidence that tumour cell surface polypodial activity appears to be the major factor in tumour invasion (Ambrose and D.Easty, 1976). However, detailed comparison between the pattern of social behaviour of tumour cells in vitro and their in vivo pathological properties can be done only within a strictly genetically defined system of inbred animals.

Some years ago, we made attempts to employ inbred Lewis rats for such studies (Veselý and Weiss, 1973). The most interesting result obtained with spontaneous in vitro and Rous sarcoma virus-transformed rat cells was marginal evidence for the significance of the occurence of a nonreciprocal contact inhibition in vitro for the prediction of in vivo invasiveness. In addition to these factual results, two guidelines emerged for future work. First, it appeared to be very difficult to obtain an experimental solid tissue tumour which would be invasive and which would disseminate. Also, the histopathological evaluation as the only criterion for the selection of experimental tumour with malignant properties resembling clinical cancers seemed to be very restrictive. Secondly, cinemicroscopic techniques used at that time were not adequate to meet the requirements of observing, registering and analysing the cell situation in vitro. In the present communication, we would like to refer to the solution to the first problem and describe our approach to the second.

TUMOUR GROWTH IN VIVO

In collaboration with Dr. V. Křen, we put together a collection of various avian sarcoma virus in vitro transformed rat cells (Veselý and colleagues, 1978), chemically in vivo induced rat tumours (Křenová and co-workers, 1978) and derivatives of spontaneously in vitro transformed rat fibroblasts (Veselý and co-workers, 1968). Using a system of genetically defined inbred and congenic rat strains (Křen and co-workers, 1973) we have further broadened the approach to the study of tumour cell growth against immunological pressure (Křen and colleagues, 1978) into a system in which the criterion of malignancy is the propensity of tumour cells to grow across histocompatibility barriers of different strength and eventually kill the host (Křen and colleagues, 1979):

1. Very low level of tumorigenicity - some virus-transformed cells grow in newborn syngeneic rats but are rejected in adults after a period of initial growth.
2. Low tumorigenicity - cells growing only in adult syngeneic rats, for example, the chemically induced FL tumour.
3. Intermediate tumorigenicity - other chemically induced rat tumours grow with varying efficiency against non-MHC immunological barriers. All virus-transformed cells progressively growing in syngeneic rats also fall into this category as they kill 100% of inoculated adult rats of the non-MHC strains.
4. High tumorigenicity - only spontaneous tumours are able to grow even across the strongest MHC barrier and kill the host.

As we have observed variants, which yielded invasive tumours, only in the population of spontaneously transformed cells, we assume that there is a linkage between the propensity of malignant ell to grow across MHC barriers and its invasiveness. Therefore we believe that the genetically defined system of immunological barriers of different strength could not only classify growth capacity and estimate potential invasiveness of malignant cells but also allow the selection of invasive cell variants.

LIGHT MICROSCOPY

When the two-dimensional model in vitro is used as an approximation to the three-dimensional situation in the organism then the disadvantage of such a simplification should be compensated by the possibility of observing the course of events between colliding normal and tumour cells with the greatest achievable resolution.

But then under the highest magnification the information on the circumstances of the situation is missing as we can observe the dynamics of the same event either at high or low magnification. This means that always one half of the operational information is not available. It causes difficulties with the evaluation of cinemicroscopic films and invalidates the significance of the data obtained in this way. Besides this, it is also important to know what may be the relationship of two events ongoing in different focal levels simultaneously, such as cellular activity in bottom and top layers in culture or cell top surface motility compared with the activity of marginal ruffles at the leading edge of the lammellipodium. We attempted to overcome these limitations of present models of light microscopes.

The basic principle of our new method (patented, 1977) aimed at improving the examination of living cells in vitro is that two objectives are used to look at a specimen from opposite sites simultaneously while each serves the other as a condenser. The application of the phase contrast techniques makes the design more complicated. It was described as a method of bidirectional image transfer in light microscopy (Malý and Veselý, 1979).

Figure 1 shows the present instrument called Analytical Biological Microscope (ABM). It is equipped with two 35mm registering cinecameras and works with two standard MEOPTA objectives (achromats: 20X, NA 0.45; 45X, NA 0.65). Minicomputer TESLA JPR 12 is used to control time-lapse filming as a rather complicated cooperation between both cameras and optical channels should be ensured. Cells to be observed can be cultured in a modified Rose chamber with two coverslips 0.17mm thick which are 0.6mm appart. Cells are located either on the bottom coverslip and examined in negative phase contrast with the objective 45X from the underside and simultaneously with the objective 20X from the overside or one cell population can be located on the bottom and the other cell population on the upper coverslip and looked at with any combination of the described objectives. Culture chamber is maintained at 37 centigrades with air curtain incubator. For filming two optical channels are separated by mechanical time sharing which is achieved by a delay of 0.2 second between the actual exposures (0.5-1 second) in upper and lower cinecameras. It makes 0.2 second time difference from simultaneity.

Fig. 1. General view of the ABM.

Four pairs of frames taken simultaneously with MEOPTA objectives 20X (left side) and 45X (right side) in negative phase contrast show the course of the collision between two cells in Fig. 2, cell division in Fig. 3, and ruffling with pinocytosis in the cell periphery in Fig. 4. With objective 45X the appropriate resolution is achieved and morhological details in the cell such as nucleoli, pinosomes, structures in the cytoplasm and surface ruffles or lobopodia are clearly discernible. On the contrary the picture presented by the objective 20X is very poor providing only the information about the number of cells in the field of view and direction of their displacement. This is caused by its inappropriate optical correction for the observation of cells on the distant coverslip across additional 0.6mm thick layer of water behind the apposite coverslip. The original objectives designed to be able to cope with such requirements were not available until now. New MEOPTA planachromats developed for ABM (32X, NA 0.8, 1.1mm coverslip, water immersion; 100X, NA 1.1, 1.1mm coverslip, water immers.) are currently tested.

In summary, three sequences of paired frames from cinemicrografic films visualizing the same cells at two different magnifications were simultaneously obtained with the method of bidirectional image transfer in light microscopy. This technique was developed to enable further progress in the study of tumour cell social behaviour in vitro. The need of a variety of invasive experimental tumours for such studies is postulated and a system for their classification and selection is referred to.

ACKNOWLEDGEMENT

The authors would like to express their gratitude to Mr. S. Vyskočil, Mr. K. Jaška and Mr. J. Černý for their technical assistance and to Dr. F. Franc for the compilation of computer program.

REFERENCES

Abercrombie, M. and E. J. Ambrose (1962). The Surface Properties of Cancer Cells: A Review. Cancer Res., 22, 525-548.
Abercrombie, M. and G. A. Dunn (1975). Adhesions of fibroblasts to substratum during contact inhibition observed by interference microscopy. Exp. Cell Res., 92, 57-62.
Abercrombie, M. and A. A. Turner (1978). Contact reactions influencing cell locomotion of a mouse sarcoma in culture. Medical Biology, 56, 299-303.
Ambrose, E. J. and Dorothy M. Easty (1976). Time-lapse Filming of Cellular Interactions within Living Tissues. III. The Role of Cell Shape. Differentiation 6, 61-70.
Czechoslovak Patent (1977), No. 178 512.
Heaysman, J. E. M. (1970). Non-reciprocal contact inhibition. Experientia, 26, 1344.
Křen, V., O. Štark, B. Frenzl, V. Bílá, D. Křenová and M. Kršiaková (1973). Rat alloantigenic systems defined through congenic strain production. Transpl. Proc., 5, 1463-1466.
Křen, V., M. Kršiaková and V. Bílá (1978). Transplantability of Ferridextran-induced LEW/CUB Tumours FLA, FLB, FLC and FL. Fol. Biol. (Praha), 24, 381-390.
Křen, V., P. Veselý, A. Jirásek, M. Kršiaková and D. Křenová (1979). A Model System for Assessment of Malignancy of Rat Tumour Cells by Evaluating the Inability to Grow Across Histocompatibity Barriers of Different Strength. Fol. Biol. (Praha). Accepted for publication.
Křenová D., M. Sladká, V. Křen and F. Bohm (1978). Ferridextran-induced Tumours in the LEW Strain, in vitro Explantation. Fol. Biol. (Praha), 24, 383-384.
Malý, M. and P. Veselý (1979). A new light microscopic method for the synchronous bidirectional illumination and viewing of living cells in different contrast modes, and/or different focal levels or magnifications. J. Microscopy. Accepted for publication.
Veselý, P., L. Donner and M. Kučerová (1968). Spontaneous malignant transformation of emryonic rat fibroblasts from an inbred Lewis strain in vitro. Fol. Biol. (Praha), 14, 409-410.
Veselý, P. and R. A. Weiss (1973). Cell locomotion and contact inhibition of normal and neoplastic rat cells. Int. J. Cancer, 11, 64-76.
Veselý, P., V. Křen, J. Wyke, A. Jirásek, M. Elleder and G. Bannikov (1978). In vitro and in vivo properties of neoplastic LEW/CUB cells. Fol. Biol. (Praha), 24, 392-394.
Wood, S. jr., R. R. Baker, R. Lewis jr. and B. Marzocchi (1967). Locomotion of cancer cells in vivo compared with normal cells. UICC Monograph Series, 6, 26-30. Springer-Verlag, Berlin.

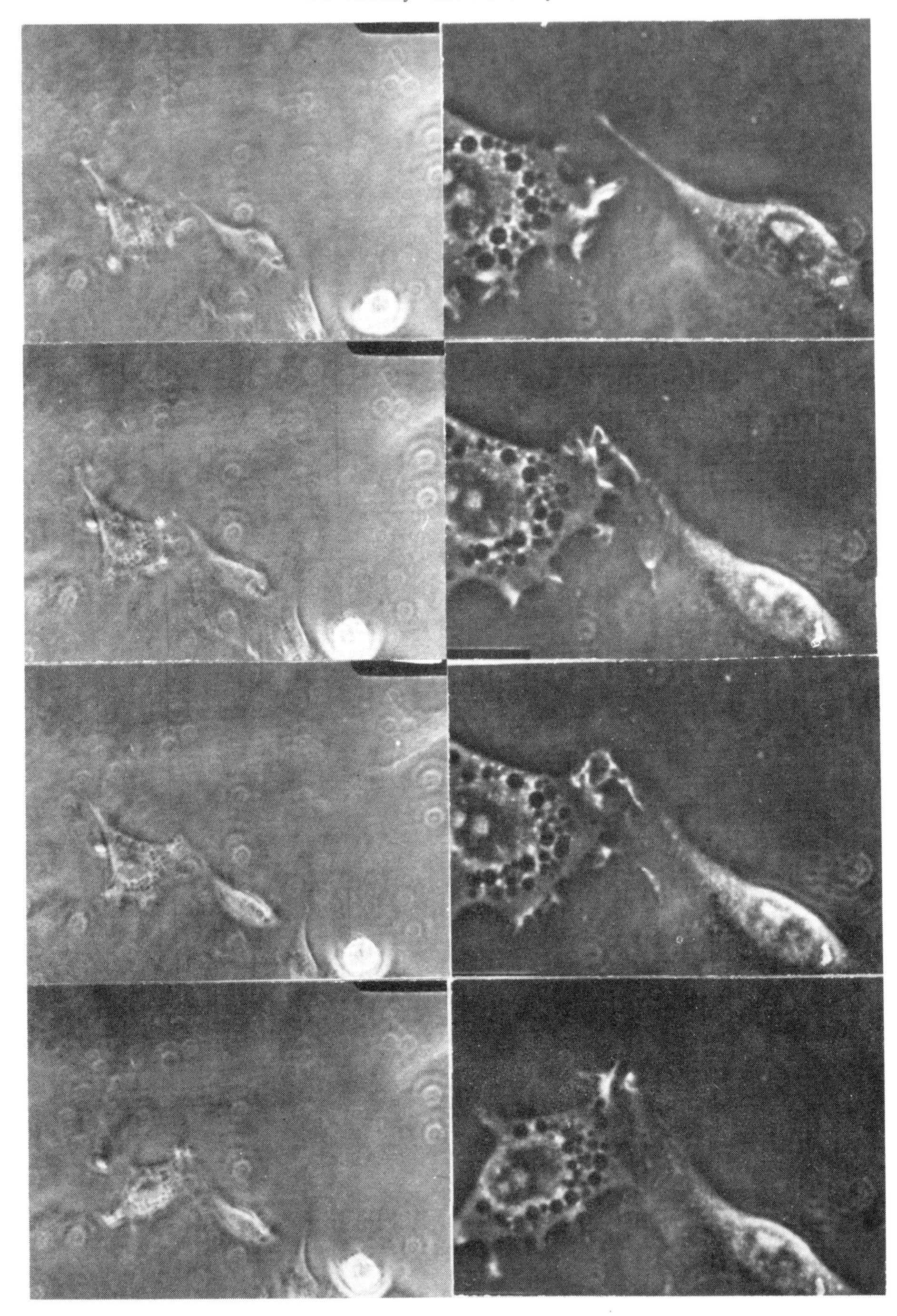

Fig. 2. Collision between two cells. Objective 20X and 45X. Sampling interval 4 minutes.

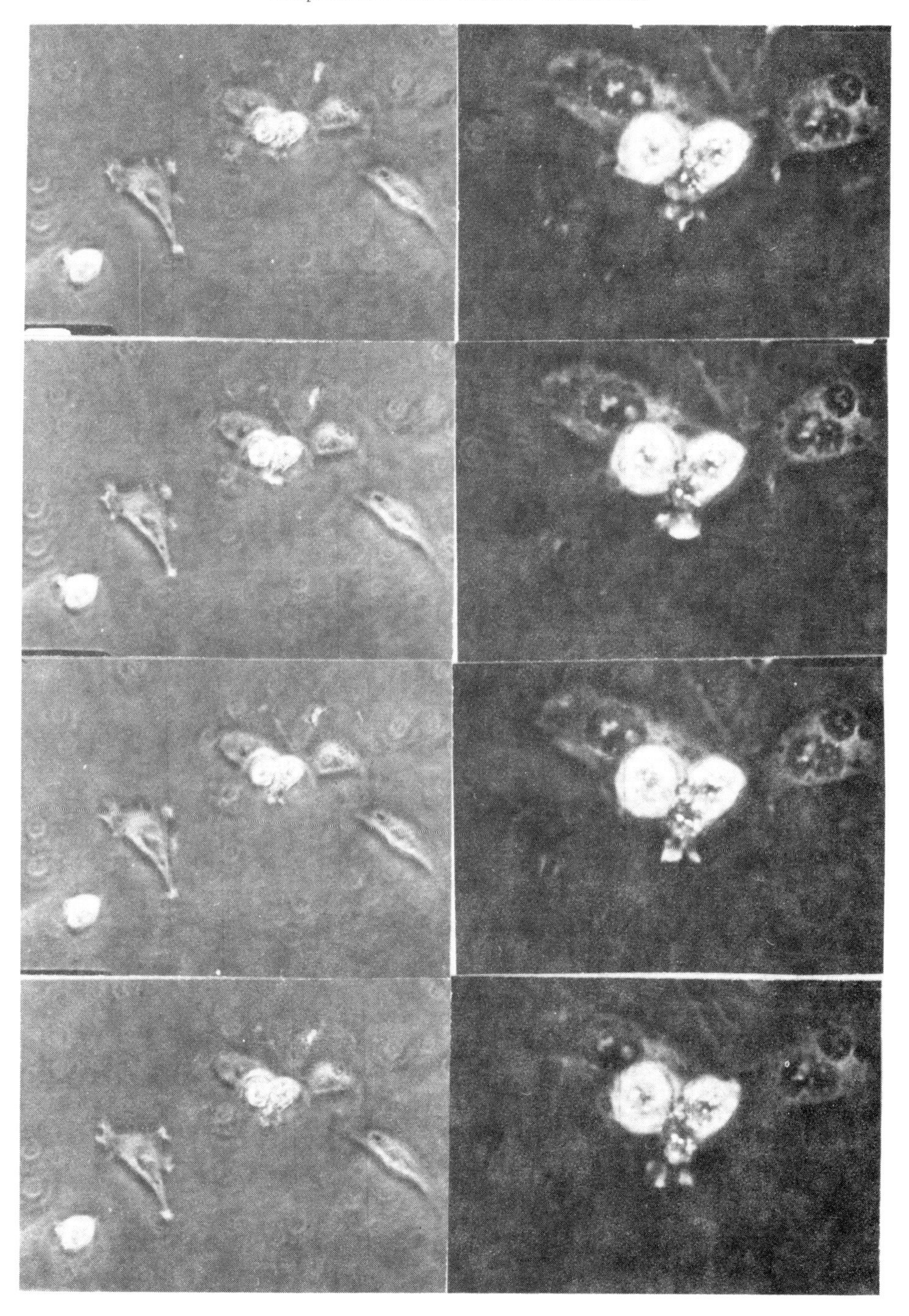

Fig. 3. Cell division. Objective 20X and 45X.
Sampling interval 10 seconds.

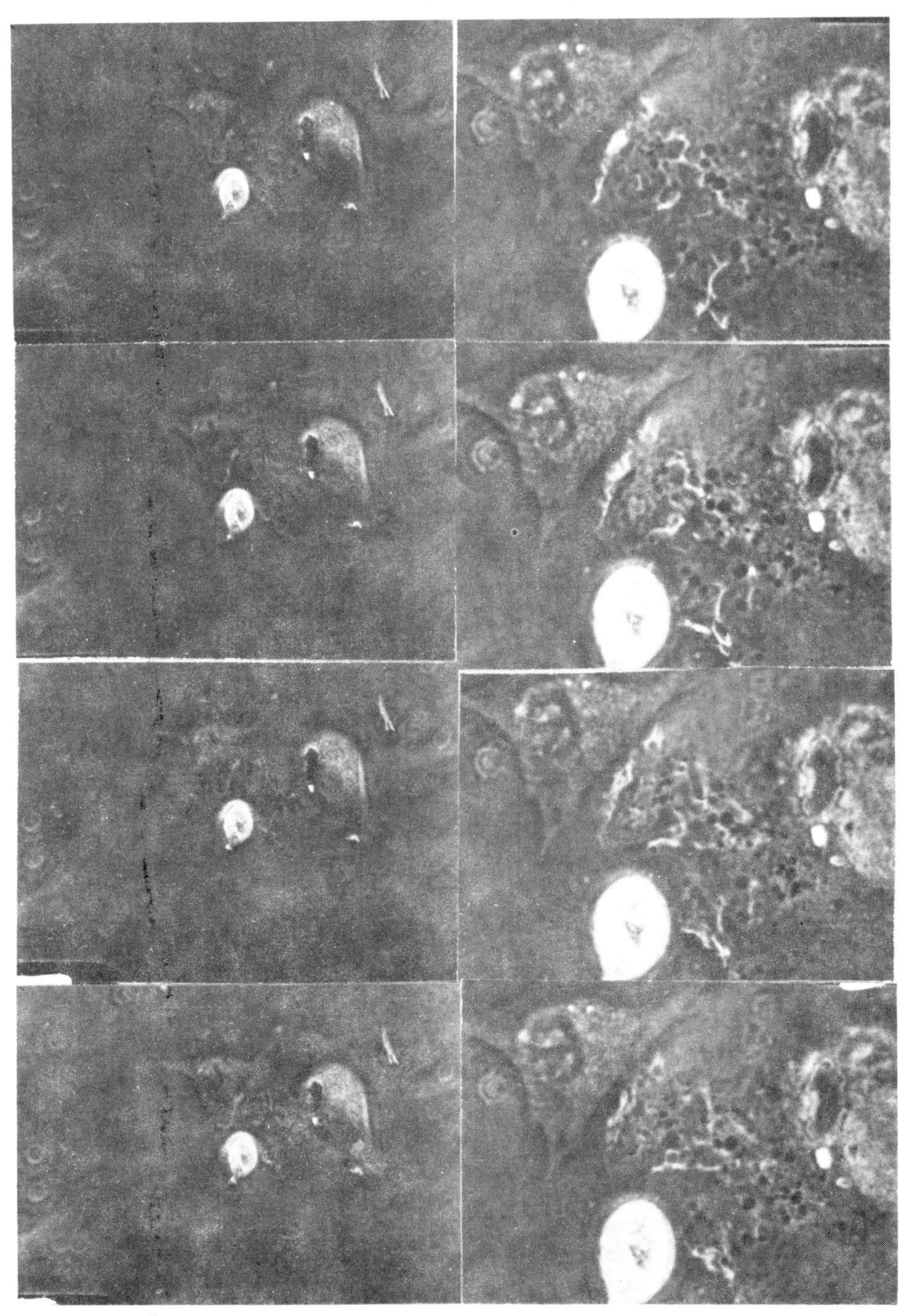

Fig. 4. Cell periphery: ruffling and pinocytosis. Objective 20X and 45X. Sampling interval 10 seconds.

Histokinetic Responses of Epithelial Cells in Histophysiologic Gradient Culture

J. Leighton, R. Tchao, W. Johnson and N. Abaza

Cancer Bioassay Laboratory, Department of Pathology, Medical College of Pennsylvania, Philadelphia, Pennsylvania, U.S.A.

ABSTRACT

In our new procedure, excised tissue is provided with a setting closely resembling that seen in nature, where membranes composed of stratified epithelial cells have their attachment and their complete metabolic exchange including oxygen on a common interface. A thin collagen diaphragm, covering one end of a shallow cylinder, is the physical substrate to which the tissue is attached, and through which it receives all nutrients. This new culture system has been evaluated with some tissues of the experimental cancer laboratory, chick embryo skin, normal rat bladder, transplantable tumors, and cell lines. Specific histokinetic responses were expressed by each of the inocula resulting, after a few days, in a spatial organization of tissue characteristic for each inoculum and similar to the tissue arrangement found in the intact animal.

KEYWORDS

Histokinetic activity; histophysiologic gradient culture; aggregate replication; neoplastic blockade.

INTRODUCTION

Histophysiologic gradient culture is a new idea in tissue culture (Leighton, 1979; Leighton, Goldberg and Stein, 1979; Leighton, Tchao and Abaza, 1979) but an old one in nature. In all mammals the interface of an epithelium and its supporting stroma is the site both of metabolic exchange with, and attachment to the interstitium. In culture the usual arrangement finds attachment at one surface, the surface of the culture vessel, and metabolic exchange on the opposite side. We have developed a procedure in which attachment and nutrition, including oxygen, occur at the same surface. Maturation characterized by the formation of special products is observed at the opposite free surface.

MATERIALS AND METHODS

Slices of cellulose sponge were cut into pieces 20 x 14 x 4 mm. In each piece two holes 1 mm apart were made with a paper punch, creating cylindrical chambers 6 mm in diameter and 4 mm deep. A thick coat of collagen (Ethicon, Somerville, N. J.) was applied to the walls of the holes. After drying and cross-linking with dilute NH_4OH, a thin membrane of collagen was affixed to one end of the cylinders, and the NH_4OH treatment was repeated. Details on the use of a collagen dispersion with cellulose sponge in tissue culture have been published (Leighton, 1973). Inocula were placed on the collagen floor of each chamber with a small amount of medium and allowed to attain attachment overnight, or attachment was accomplished by cementing the fragments in place with a chick plasma clot. By either method, when the tissue was attached to the collagen diaphragm, the sponges were inverted, placed in large Leighton tubes, and covered with a mesh of stainless steel wire. Medium was added to cover the mesh, usually 14 ml., and the tubes were incubated on a rocker at 36.5^{o}C. Inverted, the collagen floor became the roof of the chamber. Inocula attached to this roof received almost all nutrition through the collagen membrane, thus reconstituting the histophysiologic diffusion gradient of intact tissue. Living cultures were examined regularly; proliferation and movement were studied with an inverted microscope. At appropriate times cultures were fixed and studied with histologic and EM techniques. Since this is a new approach to the culture of tissues, some technical details of our model system are continuing to undergo change and improvement.

RESULTS

Pilot studies have been conducted with chick embryo skin, heart, and liver, rat bladder cancer cell line NBT II, dog kidney cell line MDCK, normal rat bladder, and a few clinical carcinomas. The medium used in all studies consisted of Waymouth's medium MB752/1 with 20% fetal bovine serum, and with antibiotics.

Eleven-day-old chick embryo skin was reduced to small fragments and placed in culture using a chick plasma clot. Extensive proliferation and migration took place. On histologic examination of two-week-old cultures the collagen diaphragms were seen to be covered with an orderly, complex tissue in distinct layers. Next to the diaphragm was a mat of loose cellular connective tissue, the dermis. The opposite surface of this connective tissue mat was covered by a stratified squamous epithelium divided into two zones. The zone adjacent to the stroma was nucleated, several cells deep, with basal cells in mitosis. With ^{3}H-TdR autoradiography this epithelial layer resembled a row of black beads. The outer free surface had several layers of keratinized cells without nuclei.

Rat bladder line NBT II, a squamous cell carcinoma, grew as a disorderly stratified epithelium with irregular prominent keratinization. Autoradiography showed that most of the labelling was in cells adjacent to the diaphragm, although labelling was also seen in cells at other strata.

In one experiment a suspension of NBT II cells was injected intravenously in 11-day-old chick embryos, and the skin removed and minced for the preparation of cultures about an hour later. After two weeks of culture the tissue had reconstituted normal skin as described above, and in addition there were scattered colonies of proliferating squamous cell carcinoma in the dermis.

Normal rat bladder provided a few surprises during 9 weeks of cultivation. Intact bladders were minced, and several fragments placed in each chamber. After one week the collagen diaphragm was completely covered with luxuriant growth. Viewed with the inverted microscope the epithelial sheets varied. Many were composed of small cells, others of very large flat cells. With time, in some chambers, the sheets of small cells covered progressively more of the diaphragm as large epithelial cells became less numerous. Fibroblastic outgrowth was never prominent. On histologic section of 2-week-old cultures the growth, characterized by small epithelial cells, consisted of a stratified epithelium that was in some areas 6 to 8 cells thick and in others lesser numbers of layers. With tritiated thymidine autoradiography, when the isotope was administered one day before fixation, the zone adjacent to the diaphragm was labelled, regardless of the number of cells in the strata. Fragments of the inoculum could be recognized here and there as collagenous masses of tissue with few residual stromal cells, none of which could be identified in the outgrowth. This microscopic picture resembled closely that designated as diffuse hyperplasia in the experimental rat bladder in situ.

Between the 7th and 9th week of culture, increasing numbers of rounded, or sausage shaped multicellular masses were seen in living cultures, arising from the areas of densely packed small epithelial cells. The masses varied in size, and some appeared to be giving rise to small secondary masses by a budding phenomenon. In histologic sections of 9-week-old cultures the nodular areas were composed of large numbers of multicellular round or oval masses, extending from the basal surface of the covering epithelium into a narrow interstitial space between the basal layer of epithelium and the collagen diaphragm. On ^{3}H-TdR autoradiography the nodules were heavily labelled.

DISCUSSION

Cells engage in coordinated movement in the course of normal maturation, reaction to injury, and in the spread of cancer. The term morphogenesis, as used to designate such movements, may be erroneous, since morphogenesis contains the idea of differentiation. I propose instead the term histokinetic activity since it applies precisely to the movements of cells that take place in the dynamic processes of changing structure and organization.

Of the fundamental problems in cancer biology that are accessible for investigation with this model, two are of particular interest in our laboratory, the biology of aggregation of carcinoma cells, and the mechanisms by which cancer cells destroy and replace normal cells. Histologic sections of conventional organ cultures show central necrosis in the explant as a common feature. In contrast, when grown in histophysiologic gradient, the tissue that undergoes necrosis is farthest from the diaphragm. It falls, leaving a viable surface where histokinetic phenomena occur freely, unhampered by masses of necrosis.

As carcinomas grow there are more and more cancer cells, and more and more nests of cells. The factors regulating the replication of aggregates of carcinoma cells are completely unknown, although there have been pilot studies on this subject in tissue culture (Leighton, 1959; Leighton and co-workers, 1960). Early models were unsatisfactory. The future of studies on the replication of aggregates may now be brighter using histophysiologic gradient procedures. This is indicated by our observations on cultures more than 2 months old, prepared with inocula of normal adult rat bladder, where multiple nodules of epithelium erupted from the basal layer of epithelium into the interstitial space and proceeded to make secondary nodules by a kind of budding. Such phenomena of growth can be modified by chemicals in the living culture, and followed with the inverted microscope. Cultures can be fixed and examined in great detail with light and electron microscopy at selected intervals.

The destructive replacement of normal tissues by cancer is poorly understood. The mechanism may sometimes be accounted for by "neoplastic blockade" (Leighton, 1968), a phenomenon that has been suggested in which tumor cells destroy normal cells by growth and movement so that they become a barrier between normal parenchymal cells and the nutritional source.

The use of tissue culture procedures as an adjunct to surgical pathology diagnosis was the hope of those pathologists and surgeons who were major contributors to the conception and birth of tissue culture in the early years of this century. That their hopes have not been realized is attributable, I think, to the failure to reproduce in culture the histophysiologic organization of epithelial membranes. Our new model system provides a context that is recognizable to the surgical pathologist, as well as being rational to the cell biologist. As we acquire experience in histophysiologic culture with some commonly occurring types of cancer, we look forward to contributing to more effective management of individual patients. When the first tissue diagnosis is made on the basis of a metastasis, and the site of origin is in doubt, we may be able to recognize in culture the organ of origin, and thereby provide the clinical oncologist with the information required in choosing a mode of treatment.

Another diagnostic problem of interest to us is dysplasia of stratified epithelium such as may be seen in the uterine cervix. We expect that in histophysiologic gradient culture, these lesions will present different patterns of growth and responses to chemicals, and thus permit us to differentiate between dysplasias that will become invasive cancer and those that will not.

ACKNOWLEDGEMENT

This investigation was supported in part by grant numbers CA 14137 and CA 17772, awarded by the National Cancer Institute, DHEW.

REFERENCES

Leighton, J. (1959). Aggregate Replication, A Factor in the Growth of Cancer. Science, 129, 466-467.

Leighton, J. (1966). Neoplastic Blockade, A New Concept in the Destruction of Normal Tissue by Cancer. In H. Katsuta (Ed.), Cancer Cells in Culture, University of Tokyo Press, Tokyo, Japan. pp. 143-156.

Leighton, J. (1973). Cell Propagation on Miscellaneous Culture Supports: Collagen-Coated Cellulose Sponge. In P. K. Kruse, Jr. and M. K. Patterson, Jr., (Eds.), Tissue Culture, Methods and Applications, Academic Press, New York City. pp. 367-372.

Leighton, J. (1979). Histophysiologic Gradient Culture - Concept, Method, Early Observations, Diagnostic Possibilities (abstract). Laboratory Investigation, 40, 268.

Leighton, J., M. Goldberg and R. Stein (1979). Histophysiologic Diffusion Gradient Culture, A Method for Three Dimensional Growth of Complex Tissues (abstract). IN VITRO, 15, 174.

Leighton, J., R. L. Kalla, J. M. Turner and R. H. Fennell (1960). Pathogenesis of Tumor Invasion II. Aggregate Replication. Cancer Res., 20, 575-586.

Leighton, J., R. Tchao and N. Abaza (1979). Histophysiologic Gradient Culture of Normal and Neoplastic Rat Bladder Epithelium as Multilayered Sheets with Both Attachment and Nutrient Exchange at the Basal Surface (abstract). Proc. Amer. Assn. for Cancer Res., 20, 2.

An Embryological Model of Non-malignant Invasion or Ingression

L. Vakaet, Chr. Vanroelen and L. Andries

Laboratorium voor Anatomie en Embryologie, Rijksuniversitair Centrum Antwerpen, Groenenborgerlaan 171, B-2020 Antwerpen, Belgium

ABSTRACT

Observations using light and electron microscopy (TEM and SEM) of normal and experimentally induced primitive streaks of the chick blastoderm allow us to propose a theory on blastoporal ingression in the fowl . The first visible sign announcing ingression is blebbing at the ventral side of the upper layer . This blebbing perforates the basal lamina ; whether this occurs by disruption or by a stop in the turnover process of the basal lamina remains an open question . The recently de-epithelized cells show an active "on spot motility", characterized by a very conspicuous blebbing . This blebbing gradually changes into the formation of filopodia without any preferential direction nor adhesion . When, during gastrulation, through the narrowing of the area of the upper layer committed to ingress, the middle layer cells contact a continuous basal lamina, their motility turns into a very active mobility, leading to the divergent migration of the mesoblast cells towards the periphery of the area pellucida .

KEYWORDS

Chick blastoderm; gastrulation; primitive streak; ingression; invasion; colonization; cell movements; basal lamina .

INTRODUCTION

By ingression we mean de-epithelization of cells, followed by their migration along the deep side of the epithelium they stem from . Colonies are cellular accumulations formed mainly by divisions of ingressed cells . There is invasion and metastasis when, moreover, ingression and colonization are accompanied by destruction. Destruction is considered to be characteristic of invasion . Ingression and colonization may indeed occur in normal systems . One of these is the gastrulating avian blastoderm where these events are part of morphogenesis . As it is, moreover, possible to experimentally evocate a secondary primitive streak, this allows us to study at will not only the way of induction of the ingression, but also the reactions within the epithelium preceding ingression of its cells .

MATERIAL

White leghorn eggs from commercial stock were used . They were incubated at 37°8 C up to the desired stage (Vakaet, 1970) .

Two steps of ingression in the avian blastoderm were studied . The first, up to stage 4, consists exclusively in de-epithelization of the cells of a large posterior part of the upper layer of the area pellucida . This phenomenon is linked to the convergence of the area bound to ingress towards the posterior midline where it culminates . There its cells build up the young primitive streak that elongates both towards the centre and towards the periphery of the area pellucida . From marking experiments, the limit of the area that will eventually reach the primitive streak and ingress is well known . The middle layer cells of the primitive streak stay in the streak up to stage 4 (Fig. 1) .

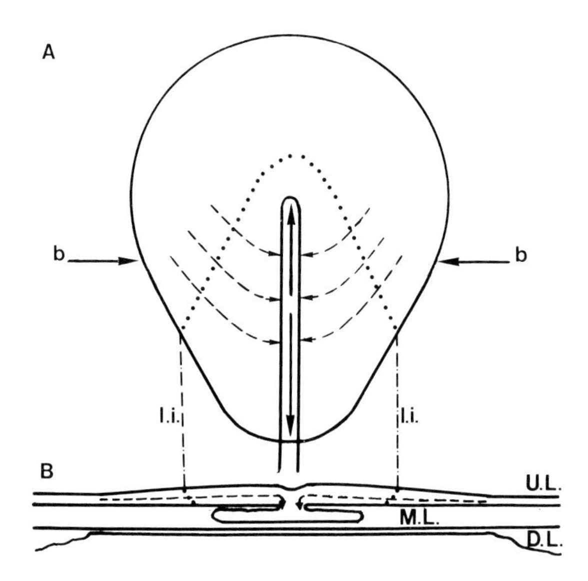

Fig. 1. Morphogenetic movements in a stage 4 chick blastoderm .
A : Outline of the area pellucida facing the dorsal aspect of the upper layer . The area is pear-shaped, due to the elongation of the primitive streak (full arrows) . The convergent movements in the upper layer are shown by dashed arrows . The limit of ingression in the upper layer is marked by l.i. . B : Section of the primitive streak through level b-b in Fig. 1 A . The disposition at this stage of the layers (the upper : UL, the middle : ML and the deep : DL) is shown and so is the limit of ingression : l.i. . The convergence and ingression movements are indicated by dashed arrows .

The second step commences when by stage 5, convergence in the upper layer comes to an end ; from that stage on, the middle layer cells of the streak start migrating towards the periphery of the area pellucida . The most peripheral of them will colonize the future area vasculosa . In stage 6 ingression is almost confined to this migratory activity (Fig. 2) .

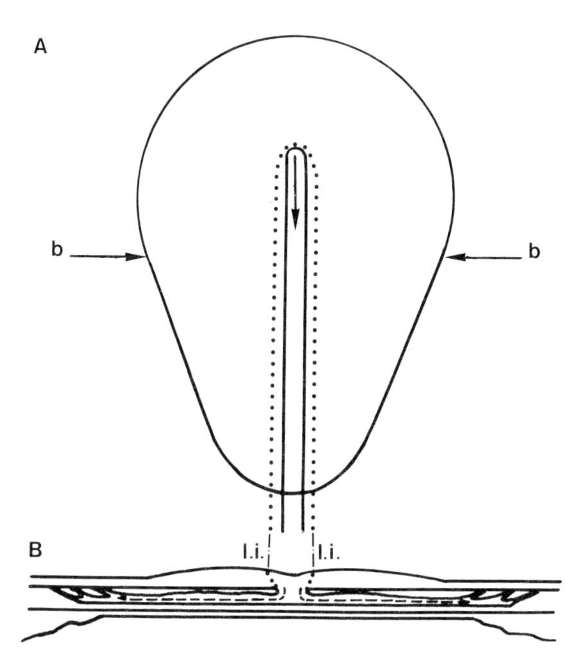

Fig. 2. Morphogenetic movements in a stage 6 chick blastoderm . A : In the primitive streak, regression of Hensen's Node has started (arrow) . The limit of ingression (dotted line, l.i.) has reached the lips of the primitive streak, and convergence has stopped . B : Section of the primitive streak through level b-b in Fig. 2 A . The l.i. is indicated in the upper layer . The middle layer cells attach long filopodia to the ventral side of the upper layer (full arrows) situated lateral to the limit of ingression . They eventually migrate very actively towards the periphery of the area pellucida (dashed arrows) .

METHODS

Microscopy

Microscopy was performed on blastoderms fixed in glutaraldehyde 1% in Millonig's buffer (1962) at a pH of 7,3 followed, after 3 rinces in the buffer solution, by postfixation in OSO_4 1% in Millonig's at the same pH . After rinsing in the buffer and in distilled water, dehydration in graded alcohols was performed .

For transmission microscopy, the blastoderms were embedded in Epon 812 . Sections 1-2μm thick for light microscopy are made on a LKB pyramitome 11800 with glass knives made on a LKB knifemaker 7800 . They have been observed in light microscopy either unstained (phasecontrast Wild M20) or stained with periodic Acid Schiff/ Toluidine blue (Kühn, 1970) .

Ultrathin sections from selected regions were made on a LKB ultratome III with glass knives, to obtain pale gold or yellow sections, or a diamond knife (F. Dehmer, Koepingstrasse 8, 8202 Bad Aibling, BRD) which permits to obtain gray sections . They are mounted on formvar coated 50 Mesh copper grids (Veco, Karel van Herleweg 22, Eerbeek, Nederland) . After staining in uranyl acetate/lead citrate after Reynolds (1963) they were coated with carbon in a Jeol JEE-4B vacuum evaporator .

In a Jeol 100B transmission electron microscope (TEM) photographs were made on Kodak 4489 electron microscopy sheet film .

For scanning electron microscopy (SEM) the blastoderms are transferred from alcohol 100° into isoamyl acetate through a graded series . After drying with a Polaron CPD apparatus using carbon dioxide as the transition solution, the dry specimens are fixed on a copper stub with colloidal silver paint (G.C.Electronics) and coated in Balzers' Union sputtering device so as to obtain a uniform gold coating of ± 200 Å . SEM observations were made with a Jeol JSM/U3 .

Induction of a Secondary Primitive Streak

For inductions of secondary primitive streaks, stage 4 to stage 5 blastoderms are used . The intervention is performed on blastoderms cultured after New (1955) . The procedure is outlined and described in Fig. 3 . The results after reincubation have been described by Vakaet (1974) .

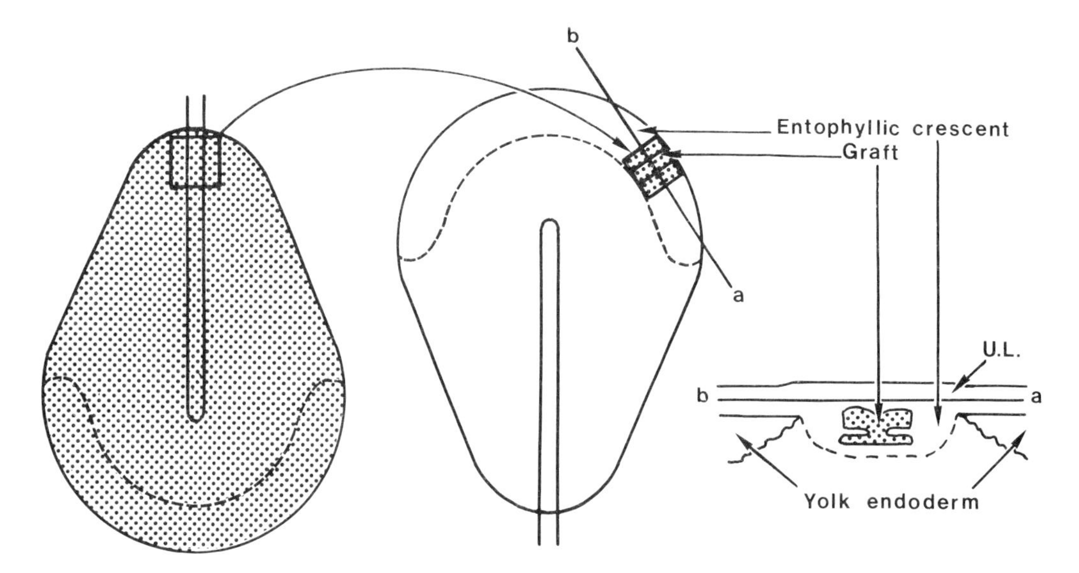

Fig. 3. Induction of a secondary primitive streak . Grafting of a posterior fragment (0,25x0,25 mm) of a primitive streak from one blastoderm (dotted) into the entophyllic crescent of another blastoderm is followed by the disappearance of the graft and an arealization due to the concentric migrating away of the cells of the graft . This is followed within 12 hours by the appearance of a well-formed primitive streak in the host upper layer, at the site of the original graft in the centre of the arealization .

RESULTS

Light Microscopy

The description of transverse sections through the anterior half of the primitive streak of stage 4 blastoderms, may summarise the light microscopic observations (Fig. 4) .

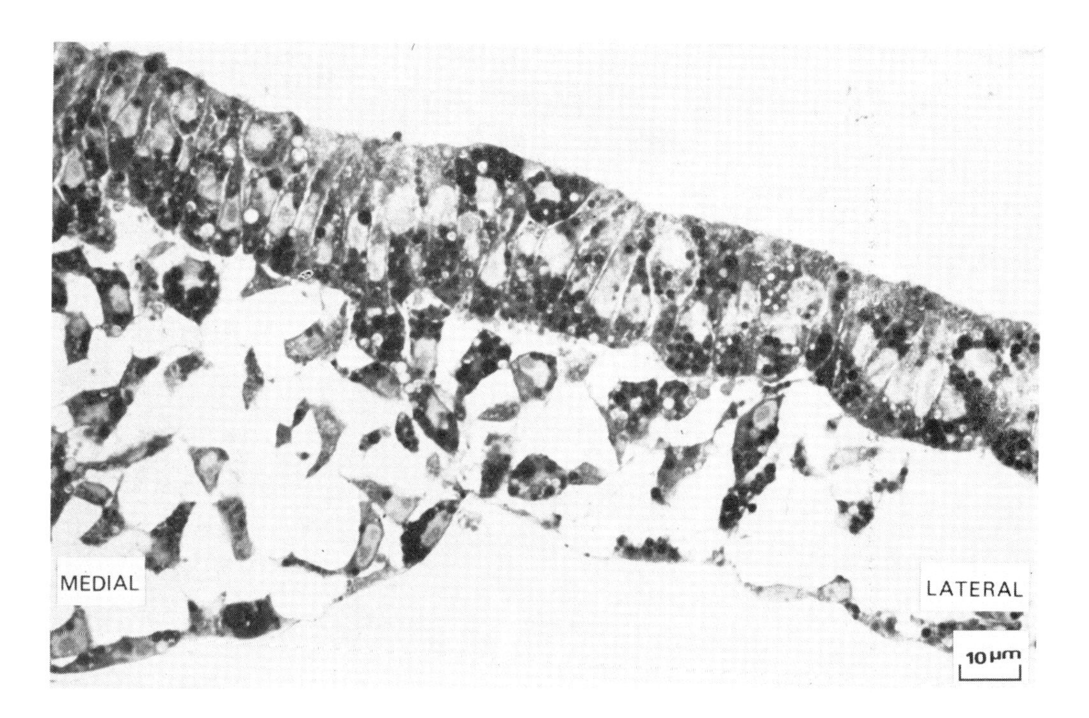

Fig. 4. Section of right side of the anterior half of a stage 4 chick primitive streak .
Leitz orthoplan ; obj. Pl. Apo 40, oil imm. . Agfachrome L50 .
For explanation, see text .

In the regions of the upper layer not disposed to ingress, the overall staining of the cells differs from that of the cells bound to leave the upper layer . The nuclei are not stained but, while the lateral part of the epithelium is blue, the medial part, next to the PS, is pale pink . We have no explanation for this signaletic staining, since after glutaraldehyde fixation the PAS reaction has no histochemical significance . This staining however has drawn our attention to another characteristic difference between the two regions, viz. the different behaviour of the ventral surface of their cells . This surface is very clearcut in the lateral area whilst in the area that will enter the primitive streak, an abundant blebbing is visible . In the middle layer cells of the streak in stage 3, this blebbing is also present . It gradually disappears and is almost absent in the mesoblast cells by stage 4 . They predominantly present slender cell extensions, but stay crowded in the streak .

Scanning Electron Microscopy

Blebbing of the upper layer is also demonstrated by SEM . Ventral views of the upper layer of a stage 4 blastoderm from which the deep layer was flushed away before fixation, show a blebbing distribution corresponding to that suggested by the study of semi-thin sections . Few blebs are visible on the upper layer that will not participate in ingression whilst they are very numerous in the area just in front and lateral to the primitive streak at this stage . On the ventral side of the middle layer cells blebs and filopodia are present at stage 4 (Fig. 5) .

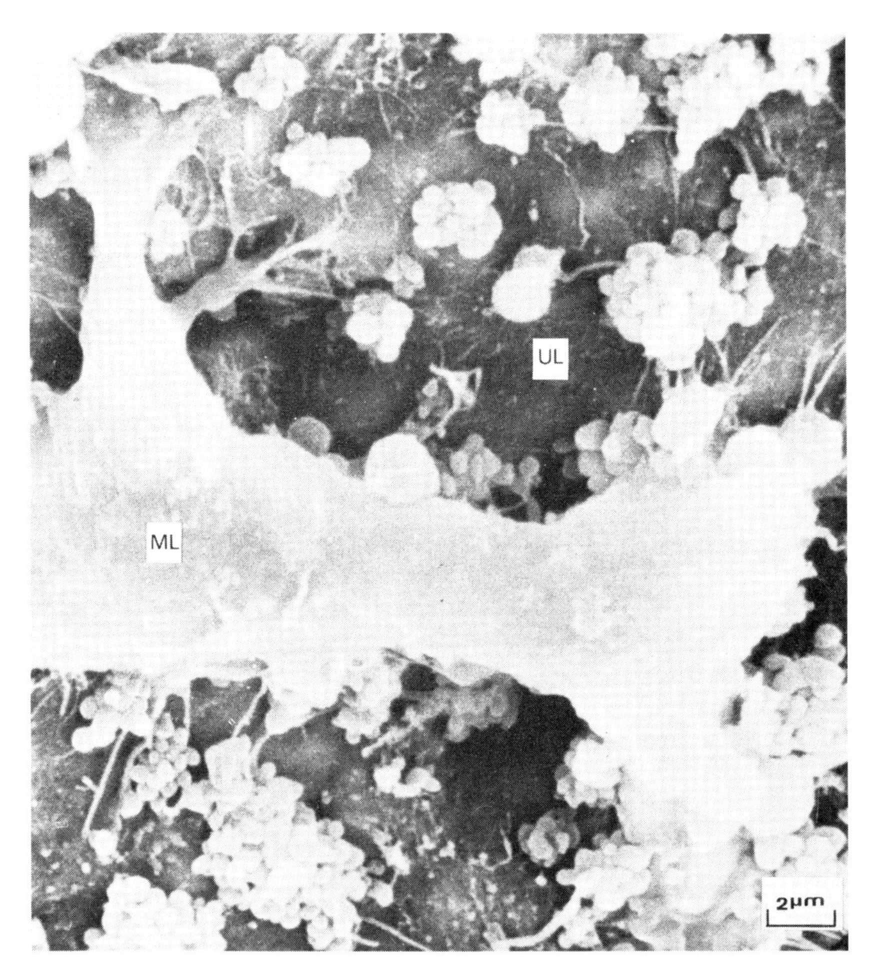

Fig. 5. SEM of a stage 4 blastoderm facing the ventral aspect of the upper layer after flushing away of the deep layer . The region shown is close to the primitive streak . Blebs are conspicuous on the upper layer and on the recently de-epithelized middle layer cells .

Stage 6 mesoblast cells and especially cells that migrate laterally exhibit very long and slender filopodia, many of which are attached to the ventral side of the upper layer (Fig. 6) .

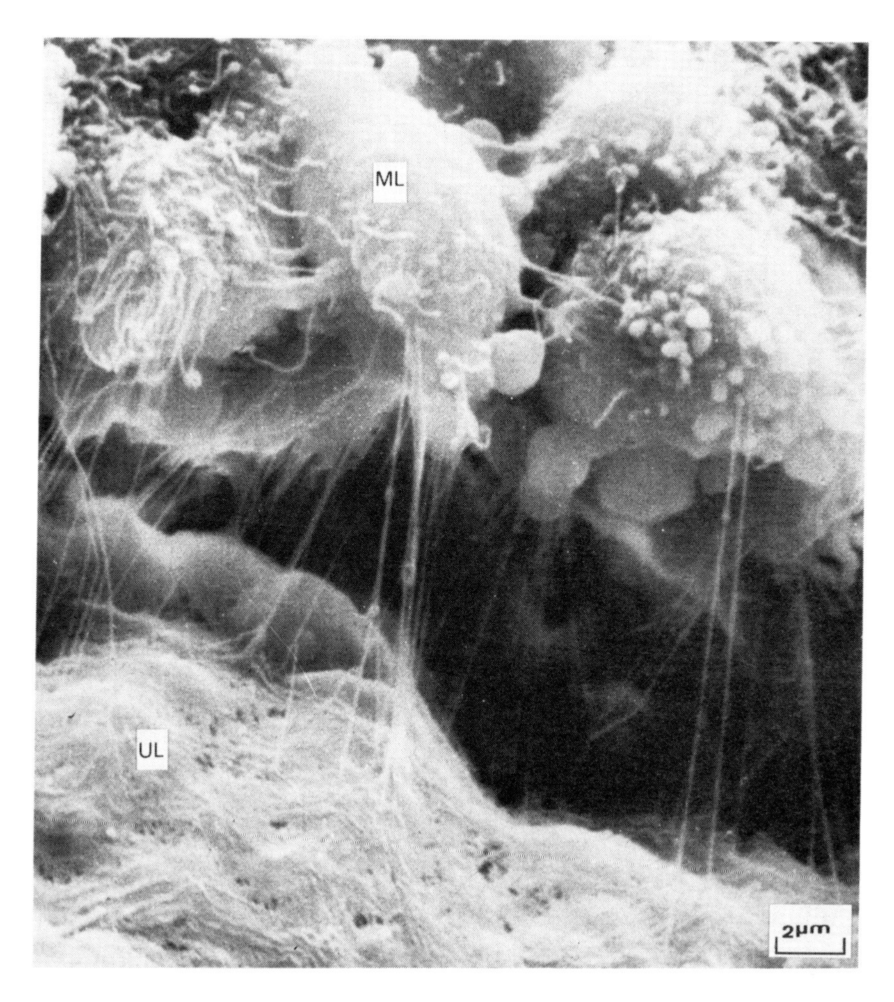

Fig. 6. SEM of a stage 6 blastoderm facing the ventral side of the upper layer and the lateral side of the migrating mesoblast sheet . No blebs are present on the upper layer, very few on the mesoblast, which is attached with numerous filopodia to the basal lamina of the upper layer .

Transmission Electron Microscopy

The nature of the blebs and their relations to the basal lamina have been studied with TEM . At stage 4, a continuous basal lamina is present in the lateral area (Fig. 7) . In the area of the upper layer bound to ingress, the basal lamina is perforated by numerous blebs (Fig. 8) . These blebs are packed with ribosomes, more densely than in any other part of the cytoplasm ; no other cellular structures are visible . These zeiotic blebs are variable in diameter (up to 2 and even 3 μm) and correspond to the structures observed with the other techniques . In the primitive streak region the basal lamina is disrupted . Parts of it are still to be found at the ventral side of cells that have reached the middle layer . On these cells, very few blebs are visible (Fig. 9) .

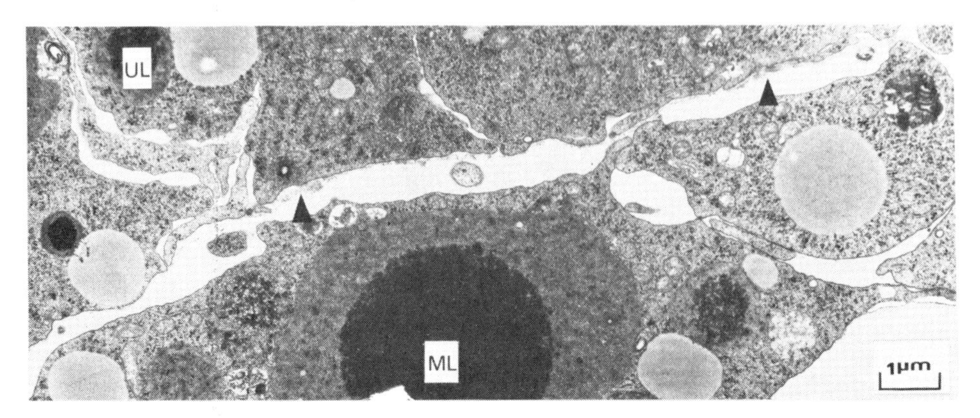

Fig. 7. TEM of a lateral part of a stage 4 blastoderm . Arrows point to the basal lamina .

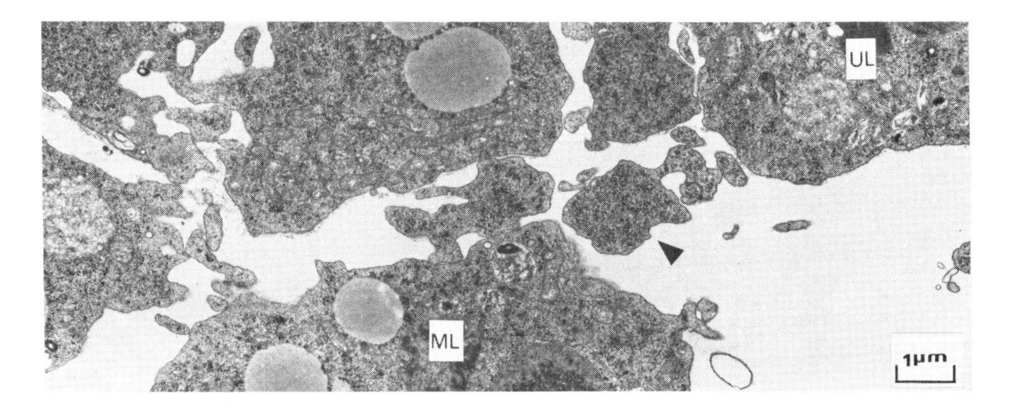

Fig. 8. TEM of a part of the UL bound to ingress . Blebs are indicated by arrow .

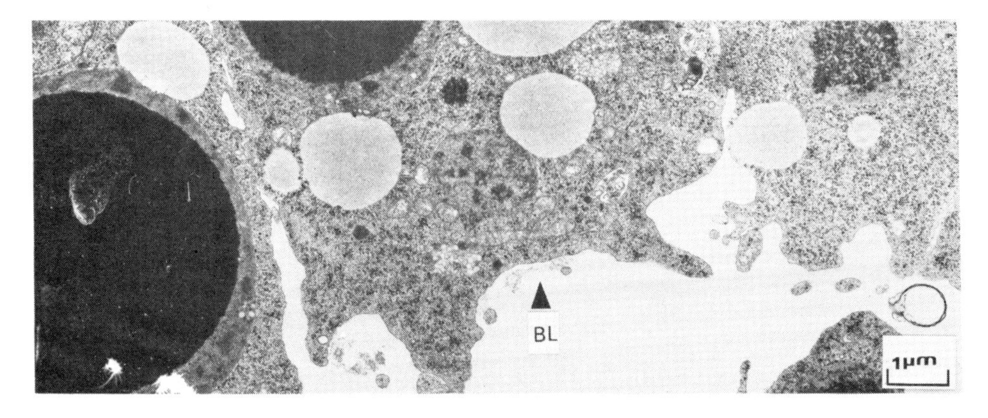

Fig. 9. TEM of recently de-epithelized middle layer cells, one of which is carrying basal lamina like material on its ventral side (arrow + BL) .

Experimental Induction of a Primitive Streak

By experimental induction of a primitive streak, we have tried to answer the question whether the interruptions in the basal lamina or the blebbing activity are the first signs of commitment of these cells to ingression . These studies show that after 5 hours, the first stage of the experimentally induced ingression are interruptions of the previously continuous basal lamina . Some blebbing activity has also started . The interruptions of the basal lamina and the blebs are to be found where grafted cells closely approach the ventral side of the upper layer (Fig. 10) .

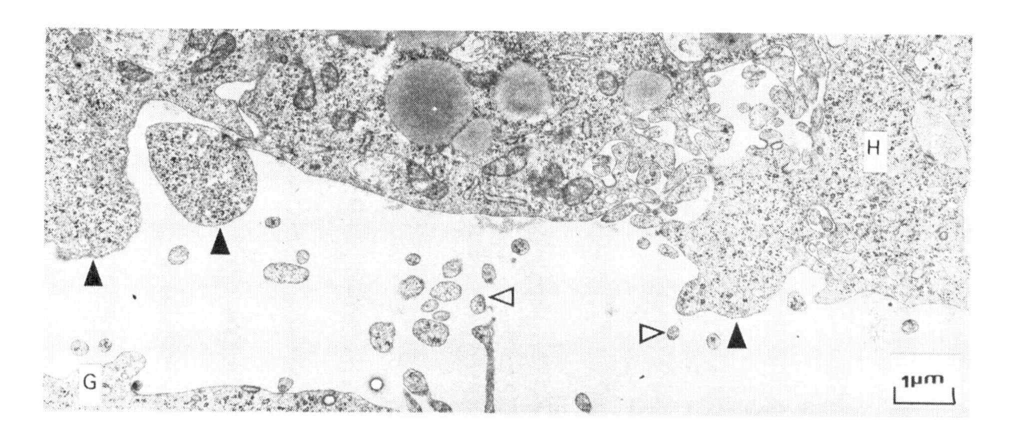

Fig. 10. TEM of the first stage of induction of a secondary primitive streak in the UL of a stage 4 chick blastoderm (Host=H) . In the region of inoculation normally no blebs occur, and the basal lamina is continuous . Five hours after grafting, many blebs have appeared (arrows), perforating the basal lamina, the remnants of which are scattered over the ventral side of the UL . Graft cells (G) have many long extensions, some of them (white arrows) approaching very closely the ventral side of the upper layer .

DISCUSSION

Our results deal essentially with the onset of blastoporal ingression in the chick blastoderm . Blebbing of cells preparing for gastrulation movements has been described by Trinkaus (1976) in *Fundulus* and Van Haarlem (1979) in Fishes *(Notobranchius Korthausae)* . According to these authors, blebbing occurs in stationary cells, while long cell extensions are present in locomotory cells . Our observations in the chick mesoblast preparing for ingression fit in with this description . Indeed, blebbing at the ventral side is the first sign by which the upper layer cells that are committed to ingress are recognizable .

TEM demonstrates that these blebs are zeiotic blebs, packed with ribosomes . The mechanism of the protrusion of these short cell extensions is unknown . Also unknown is the relation between the disruption of the basal lamina and the blebbing of these cells preparing to leave the upper layer . The disruption might be mechanical or it might be due to a change in the turnover process of the basal lamina . The change in the basal lamina might precede and even evoke the blebbing, as the experimental evidence shows (Fig. 10) .

The blebbing of the upper layer cells characteristically is an "on spot motility" . Although these cells are caught up in the morphogenetic movement towards the primitive streak (Fig. 1), they seem not to move in respect to one another . We have shown that at stage 4 this blebbing gradually changes into the formation of longer cell extensions (Fig. 5) . This change in motility does not however bring about by itself displacement of these cells . Filopodia extend parallel to the upper layer of the blastoderm, without any preferential direction . It is worthwhile to note that only small lamellipodia are present and that in no case ruffling membranes have been observed .

We observe that by stage 5 suddenly the motility of the de-epithelized cells turns into a very active mobility . By that time, through the narrowing of the area of the upper layer bound to ingress, the de-epithelized cells contact a whole basal lamina . We believe that this contact is the prerequisite for mesoblast cell movements, and is indeed the trigger for their migratory activity towards the periphery of the area pellucida .

How can we understand this divergence of the mesoblast cells, once their mobility started ? An explanation might be mutual contact inhibition of the mesoblast cells . Indeed, if the intrinsic mobility of these cells is random, and if their source is the primitive streak, the only direction mutually contact inhibiting cells can take is a divergent one . Trelstad, Hay and Revel (1967) already put forward the hypothesis of contact inhibition to explain the lateral migration of the mesoblast cells during avian gastrulation . The mesoblast cells are not inhibited by contacts with the basal lamina of the upper layer, but probably use it to establish anchoring sites during their migration (Fig. 6) . Mutual contact inhibition on the one hand and adhesion to a basal lamina not disrupted by blebbing of its overlying upper layer cells on the other hand may be two of the factors responsible for the final placement of the mesoblast cells in the chick blastoderm . The most peripheral of these mesoblast cells halt at the inner margin of the area opaca . They will colonize this region to form the area vasculosa .

ACKNOWLEDGEMENTS

We gratefully acknowledge the excellent technical assistance provided by Mr F. De Bruyn (drawing), Mrs E. Haest-Van Nueten (experiments), Mrs A. Seghers-Geldmeyer and Mrs V. Gonthier-Van der Stock (SEM), Mr G. Van den Broeck and Mr Ch. De Schepper (TEM), Mrs N. Van den hende-Bol (typography) and Mr J. Van Ermenghem (photography) .

REFERENCES

Kühn, C. (1970). Application de techniques de colorations utilisées en microscopie optique, aux coupes sémi-fines de microscopie électronique . *Das Medizinische Lab.*, 23, 204-216 .

Luft, J.H. (1961). Improvements in epoxy resin embedding methods . *J.Biophys. Biochem.Cytol.*, 9, 409-414 .

Reynolds, E.S. (1963). The use of lead citrate at high pH as an electron-opaque stain in electron microscopy . *J.Cell Biol.*, 17, 208-212 .

Trelstad, R.L., E.D. Hay and J.P. Revel (1967). Cell contact during early morphogenesis in the chick embryo . *Dev.Biol.*, 16, 78-106 .

Trinkaus, J.P. (1976). On the mechanisms of metazoan cell movements . In J. Poste and G.L. Nicholson (Eds.), *The Cell Surface*, North-Holland P.C., Amsterdam . pp. 225-329 .

Vakaet, L. (1970). Cinephotomicrographic investigations of gastrulation in the chick blastoderm . *Arch.Biol.*, 81, 387-426 .

Vakaet, L. (1973). Inductions par le noeud postérieur de la ligne primitive des oiseaux . *C.R.Soc.Biol.*, 167, 1053 .

Van Haarlem, R. (1979). Contact inhibition of overlapping : one of the factors involved in deep cell epiboly of *Notobranchius Korthausae* . *Dev.Biol.*, 70, 171-179 .

Cell Interactions and Invasion *in vitro*

The Interaction of Normal and Malignant Rat Liver Epithelial in Culture

P. T. Iype

Chemical Carcinogenesis Program, Frederick Cancer Research Center, Frederick, Maryland 21701, U.S.A.

ABSTRACT

Control of proliferation of rat liver epithelial cells which exhibit density-dependent inhibition was studied in both normal and transformed cells under different experimental conditions. The proliferation of normal rat hepatic cells is limited not only by the surface available for cellular spreading, but also by putative proliferation inhibitors produced by these cells. In contrast, the proliferation of transformed hepatic epithelial cells is limited mainly by the growth surface. The fate of cells plated over a normal cell monolayer was studied by time-lapse photography.

The normal cells merged with the monolayer and did not divide, but proliferation of hepatoma cells was not affected by the monolayer base of normal liver cells.

The cell-cycle-times were also determined from the time-lapse film for the normal cells which were placed in the center or on the periphery of a colony. The cell-cycle-time was considerably longer for the peripheral cells (27.8 $\pm$ 8.0 hr) compared with the central cells (16.5 $\pm$ 3.2 hr). Out of the thirteen cells in the colony at the start of the experiment, ten had undergone cell division during the 60-hr observation period. The progeny of three cells did not divide again, but few cells divided twice during this period.

The progeny of any particular cell did not divide synchronously, which suggests that the proliferation of each of these non-malignant cells is controlled by its extracellular micro-environment.

KEYWORDS

Proliferation control; epithelial cells; liver inhibitors; time-lapse photography; density-dependent inhibition.

INTRODUCTION

Fibroblastic cell division in culture is limited mainly by the availability of essential growth factors in the culture medium, whereas epithelial cell division is limited primarily by the amount of surface available for cell attachment

(Dulbecco and Elington, 1973). It is not clear whether density-dependent inhibition of division (Stoker and Rubin, 1967) in epithelial cells is attained when cells are in intimate contact (Loewenstein, 1979), or when nutritional and other positive growth factors are removed from the medium (Dulbecco, 1970; Stoker, 1973; Holley, 1975), or when negative factors (Stoker, 1974; Whittenberger and Glaser, 1978; Iype and Allen, 1979; McMahon and Iype, 1979) are produced by the monolayer of epithelial cells. Control of epithelial cell proliferation and its alteration in cancer cells are important yet unresolved questions in cancer research. We have studied some of these problems using well-characterized rat liver epithelial cells (Iype, 1971, 1974; Iype, Baldwin and Glaves, 1972; Iype, Allen and Pillinger, 1975; Allen, Iype and Murphy, 1976) and hepatoma cells in culture under different experimental conditions.

MATERIALS AND METHODS

A non-malignant rat liver epithelial cell line (NRL 11) and a spontaneously transformed cell line (NRLST) derived from NRL 11 were maintained routinely as monolayer cultures in Hams F10 (Ham, 1963) medium supplemented with 10% fetal calf serum. In some experiments, a highly malignant cell line (HL5) derived from a primary hepatocellular carcinoma induced _in vivo_ by oral administration of 4-dimethylaminoazobenzene was also used. These cell lines were grown on plastic Petri dishes (Falcon) maintained at 37°C in humidity cabinets with a gas phase of 5% carbon dioxide in air. The monolayers were subcultured weekly with 0.05% trypsin.

EXPERIMENTAL PROCEDURES

The control cell line, NRL 11, was plated at a density of 4×10^3 cells/cm^2 and maintained in culture with a change of medium on the day after plating and every 3 days thereafter. Confluent monolayers of approximately $6\text{-}7 \times 10^4$ cells/cm^2 were obtained after 7 days. The ability of these live base-layers to sustain growth of the normal, as well as the transformed, liver cells was investigated as described previously (Iype and Allen, 1979). For the current experiments, confluent monolayers of NRL 11 cells were prepared over half the surface area of the Petri dish by plating the cells onto dishes maintained at an angle. The cells under study were then plated over the entire dish (20-40 cells/cm^2) so that these cells would fall on both the monolayer and the clear plastic surface. The cultures were then maintained horizontally for 8-10 days with additional changes of culture medium, either with or without a partially purified proliferation inhibitor (McMahon and Iype, 1979) isolated from adult rat liver. At the conclusion of these experiments. the dishes were washed with saline, fixed in methanol, and stained with Giemsa. For time-lapse photography, the cells were plated in plastic (Falcon) T-30 flasks, placed in a tight-fitting jacket with circulating water at 37°C, and fixed on a Zeiss phase-contrast inverted microscope. The cells under study were plated at low cell densities, either on plastic surface or over a monolayer base of normal cells, and photographed at 3-min intervals.

RESULTS AND DISCUSSION

Experiments using time-lapse photography have shown that normal rat liver epithelial cells attach to and spread on the plastic substratum. When divisions occur, the cells in the entire colony slide towards the perimeter, thereby expanding the size of the colony. Neither individual cells nor the cell colonies show any locomotor movements. Despite the occurrence of desmosomal connection between cultured normal liver epithelial cells (Iype, Allen, and Pillinger, 1975), the cells round up and lose contact with the adjacent cells (Fig. 1.1) before division. Moreover, cell division also occurs at the central area of the colonies where the cells are in contact on all sides, suggesting that the intimate contact between epithelial cells does not inhibit their proliferation. The cycling times for normal cells placed in the center or on the periphery of a colony were determined from the time-lapse film. The duration of the cell cycle was considerably longer for the peripheral cells (27.8 ± 8.0 hr) compared with the central cells (16.5 ± 3.2 hr). The period between the rounding up of the cells, mitosis, and reattachment was remarkably constant for normal cells (33.4 ± 1 min), and no difference was observed between peripheral and central cells.

Of the thirteen cells in a colony at the onset of the experiment, ten had undergone cell division during the 60-hr observation period. The progeny of three cells did not divide again and two other cells divided twice during this period. Moreover, no progeny of any particular cell were able to divide synchronously, which suggests that in these monolayer cells proliferation is controlled by the extracellular micro-environment.

The transformed cells within a colony (Fig 1.2) did show considerable undulating movement. The shape of the whole colony and the position of any particular cell within a colony were substantially altered within a short period (Fig. 1.2). Therefore, the duration of the cell cycle for these cells could not be assessed from the time-lapse film taken at 3-min intervals. The time for rounding up and reattachment was much longer (88 ± 21) than that of the normal cells. The delay for reattachment to the substratum may be attributed to the altered properties of these cells, such as their ability to divide in an anchorage-independent condition (Allen and Iype, unpublished data) and also the increased mobility observed in this study.

Recently, we showed that normal rat liver cells failed to form colonies over a base-layer of living parent cells, whereas transformed cells formed distinct colonies (Iype and Allen, 1979). On the other hand, normal cells did form regular colonies when the base-layer was pre-fixed with glutaraldehyde. It was not clear from that study whether the living base-layer killed the few normal cells plated over it, or whether these cells merged with the monolayer base and were unable to divide. Using time-lapse photography it was shown that the latter possibility was correct (Fig. 1.3; 1.4). The plated cells settled over the base-layer and slowly merged with it as part of the base-layer on the substratum. As with the original cells of which the baselayer was composed, the proliferation of the added cells was also inhibited in this condition of confluency.

The transformed liver cells also attached to the substratum (Fig 1.5) but moved in between the normal cell monolayer. The movement was not restricted to one direction, and some of the normal cells in contact with the leading edge of the transformed cells were killed, as evidenced by their detachment and rupture. Expanding colonies from the dividing transformed cells could be seen (Fig. 1.6). They resembled the invading colonies observed earlier in the fixed preparations (Iype and Allen, 1979).

It became clear from the time-lapse study that the added cells did attach to the substratum, and the proliferation of normal cells was inhibited by the confluent monolayer base. This is consistent with Dulbecco and Elington's (1973) observation that growth-surface availability does control cell division in normal epithelial cells. Having established that contact between epithelial cells does not stop cell division (Fig. 1.1) in colonial growth conditions where excess growth surface is still available for the spreading of the colonies, other mechanisms were sought to explain the phenomenon of density-dependent inhibition of division. Stoker (1973) has suggested the existence of a diffusion boundary layer in the culture medium close to the cell surface with a low diffusion potential which limits the availability of nutrients to the cells. This barrier could also operate in the opposite manner by keeping out inhibitory molecules released by the cells (Stoker, 1973). Recently, Whittenberger and Glaser (1978) have shown that cell saturation density is not determined by a diffusion-limited process, using culture media with different viscosities. We found that once a confluent monolayer was formed on the 7th day, nutrient and serum growth factors supplied during regular medium changes for an additional 8-10 days did not increase the saturation density of the normal liver cells. It is therefore unlikely that a lack of exogenous growth promoters could cause density-dependent inhibition of division.

We previously suggested (Iype and Allen, 1979) the possible existence of proliferation inhibitors produced by the cell monolayer and showed colony formation of added normal cells on pre-fixed gluteraldehyde but not on living monolayers.

When a dish partially covered with a live base-layer was used, in the present experiments the added normal cells again failed to form colonies on the base-layer but did produce colonies on the clear area in the same dish (Fig. 2A). This suggests that a certain threshold concentration of the putative proliferation inhibitors in the micro-environment of the cell is necessary to inhibit cell division. In contrast, the transformed cells grew in both areas (Fig. 2C) although the colonies over the base layer were rather small. We have recently reported (McMahon and Iype, 1979) the isolation and partial purification of a rat liver fraction which specifically inhibits the proliferation of normal rat liver cells. When a very low concentration (25 μg/ml) of this fraction which is not cytotoxic was added to the cells in the full growth medium containing fetal calf serum, the normal cells did not form colonies even in the clear area of the dish (Fig. 2C). Proliferation of transformed cells, on the other hand, was unaffected by this factor (Fig. 2D). Work is in progress to concentrate and purify this proliferation inhibitor from normal liver cells and the "conditioned" medium from the normal liver cells, and to study its mechanism of action in controlling cell division.

It has been shown that proliferation of normal rat liver epithelial cells is limited not only by the available surface for attachment and spreading, but also by negative signals produced by the epithelial cells. On the other hand, proliferation of transformed liver epithelial cells is dependent primarily on the growth surface and is unaffected or minimally affected by the proliferation inhibitors of normal cells.

Fig. 1. Time-lapse microphotographs of rat liver epithelial cells taken at 3-min intervals. The chronological sequence of the pictures starts at the bottom.

Fig. 1.1. A normal cell colony on plastic substratum with a dividing cell in the center of the colony.

Fig. 1.2. A transformed cell colony on plastic substratum note the change in the morphology of the colony within 30 min.

Fig. 1.3. Added normal cells (outlined) settling over the live monolayer base.

Fig. 1.4. Added normal cells (outlined) merging with the monolayer base.

Fig. 1.5. Added transformed cells (outlined) merging with the monolayer base.

Fig. 1.6. Transformed cell colony expanding within the monolayer base.

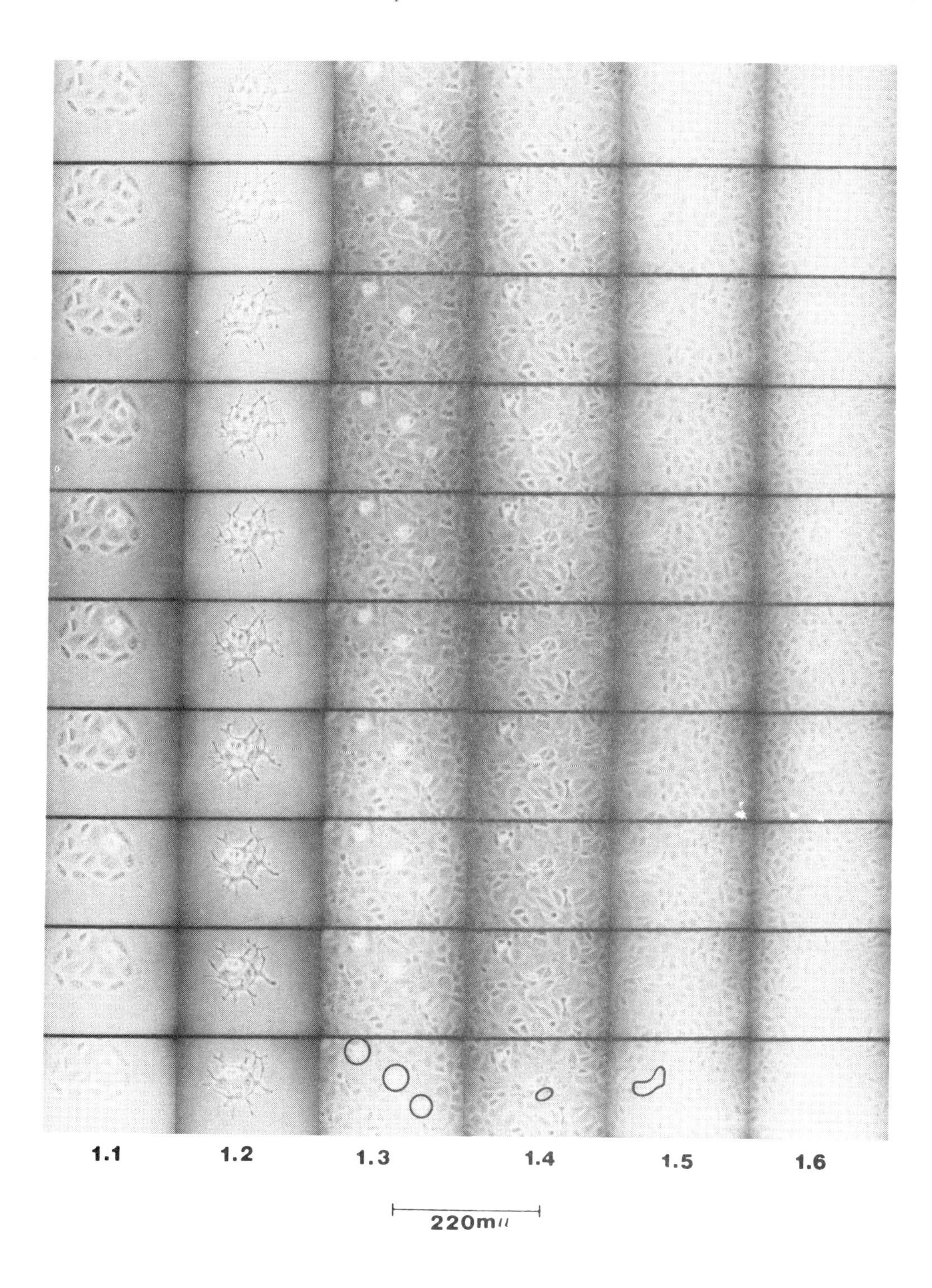
1.1
1.2
1.3
1.4
1.5
1.6
220mμ

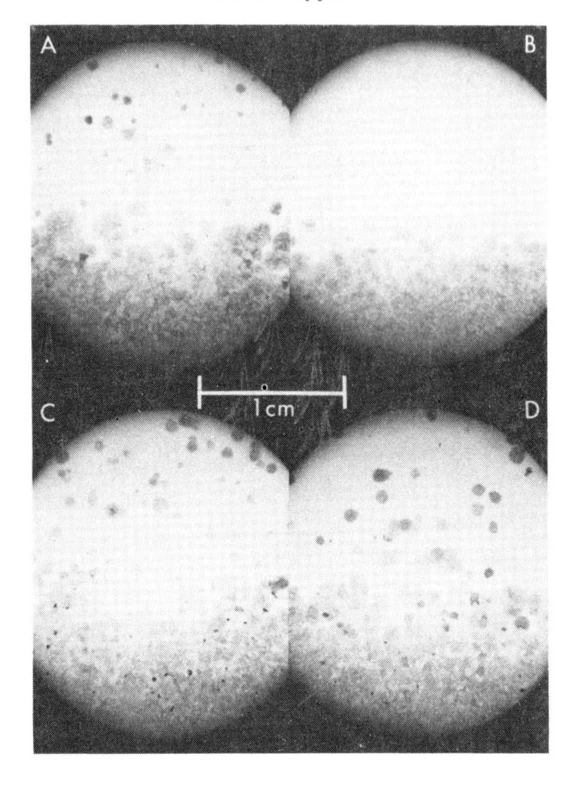

Fig. 2. Growth pattern of added rat liver epithelial cells on plastic dishes partially covered with a live monolayer base.
Fig. 2A. Normal cells.
Fig. 2B. Normal cells plus proliferation inhibitor from rat liver.
Fig. 2C. Transformed cells.
Fig. 2D. Transformed cells plus proliferation inhibitor from rat liver.

ACKNOWLEDGEMENTS

This work was supported by Contract N01-CO-75380 with the National Cancer Institute, NIH, Bethesda, Maryland 20205. The author is grateful to Mr. Albert Herring and Mrs. Carol Pontzer for technical assistance.

REFERENCES

Allen, T.D., P.T. Iype, and M.J. Murphy (1976). The surface morphology of normal and malignant rat liver epithelial cells in culture. In Vitro, 12, 837-844.
Dulbecco, R. (1970). Topoinhibition and serum requirement of transformed and untransformed cells. Nature, 227, 802-806.
Dulbecco, R., and J. Elington (1973). Conditions limiting multiplication of fibroblastic and epithelial cells in dense cultures. Nature, 246, 197-199.
Holley, R.W. (1975). Control of growth of mammalian cells in culture. Nature, 258, 487-490.

Iype, P.T. (1971). Cultures from adult rat liver cells. I. Establishment of monolayer cell cultures from normal liver. J. Cell Physiol., 78, 281-288.

Iype, P.T., R.W. Baldwin, and D. Glaves (1972). Cultures from adult rat liver cells. II. Demonstration of organ-specific cell surface antigens on cultured cells from normal liver. Br. J. Cancer, 26, 6-9.

Iype, P.T. (1974). Transformation of epithelial cells. Excerpta Med. Int. Congr. Ser. No. 350, Chem. and Viral Oncogenesis, 2, 107-112.

Iype, P.T., T.D. Allen, and D.J. Pillinger (1975). Certain aspects of chemical carcinogenesis in vitro using adult rat liver cells. In L.E. Gerschen and E.B. Thompson (Ed.), Gene Expression and Carcinogenesis in Cultured Liver, Academic Press, New York, pp. 425-440.

Iype, P.T., and T.D. Allen (1979). Identification of transformed rat liver epithelial cells in culture. Cancer Lett., 6, 27-32.

Loewenstein, W.R. (1979). Junctional intercellular communication and the control of growth. Biochem. et Biophys. Acta, 560, 1-65.

McMahon, J.B., and P.T. Iype (1979). Demonstration of a liver constituent having a differential effect on the proliferation of normal and malignant liver cells. Proceedings of American Association of Cancer Research, 20, 236.

Stoker, M.G.P., and H. Rubin (1967). Density-dependent inhibition of cell growth in culture. Nature, 215, 171-172.

Stoker, M.G.P. (1973). Role of diffusion boundary layers in contact inhibition of growth. Nature, 246, 200-203.

Stoker, M.G.P. (1974). Signals and switches: a summary. In B. Clarkson and R. Baserga (Ed.), Controls of Proliferation in Animal Cells, Cold Spring Harbor, pp. 1009-1013.

Whittenberger, B., and L. Glaser (1978). Cell saturation density is not determined by a diffusion-limited process. Nature, 272, 821-823.

Methods for Morphological and Biochemical Analysis of Invasion *in vitro*

M. Mareel, E. Bruyneel, G. De Bruyne and C. Dragonetti

Department of Experimental Cancerology, Clinic of Radiotherapy and Nuclear Medicine, Academic Hospital, De Pintelaan 135, B-9000 Ghent, Belgium

ABSTRACT

Invasiveness of malignant cells is demonstrated in vitro using three-dimensional shaker cultures of tissue fragments with aggregates of malignant cells. Both tissues are allowed to adhere to each other on a non-adhesive substrate. This method is used to examine the invasiveness of various cell lines with the aim of defining their malignancy. Cellular activities, which are presumed to be involved in invasion are studied : 1) Adhesion of malignant cells to the host tissue; 2) Destruction of the host tissue by invading cells; 3) Phagocytosis of material from the host by the malignant cells.

KEYWORDS

Invasion; malignant cells; host tissue; in vitro; three-dimensional culture; adhesion; destruction; phagocytosis.

INTRODUCTION

Invasion in vitro is a simplification of invasion in vivo. It facilitates analysis, but it holds the risk of producing results that are not applicable to invasion of malignant cells from tumours developing inside the organism. We try to limit that risk by keeping the in vitro models as close as possible to the situation in vivo on the basis of theoretical considerations and by doing comparative in vivo/ in vitro experiments (Mareel, Kint and Meyvisch, 1979; Meyvisch and Mareel, 1979).

TECHNICAL ASPECTS

Figure 1 presents a general outline of the in vitro method. This method consists of the confrontation of fragments from normal tissues, embryonic or adult (further called host tissue) with malignant cells, in three-dimensional culture.

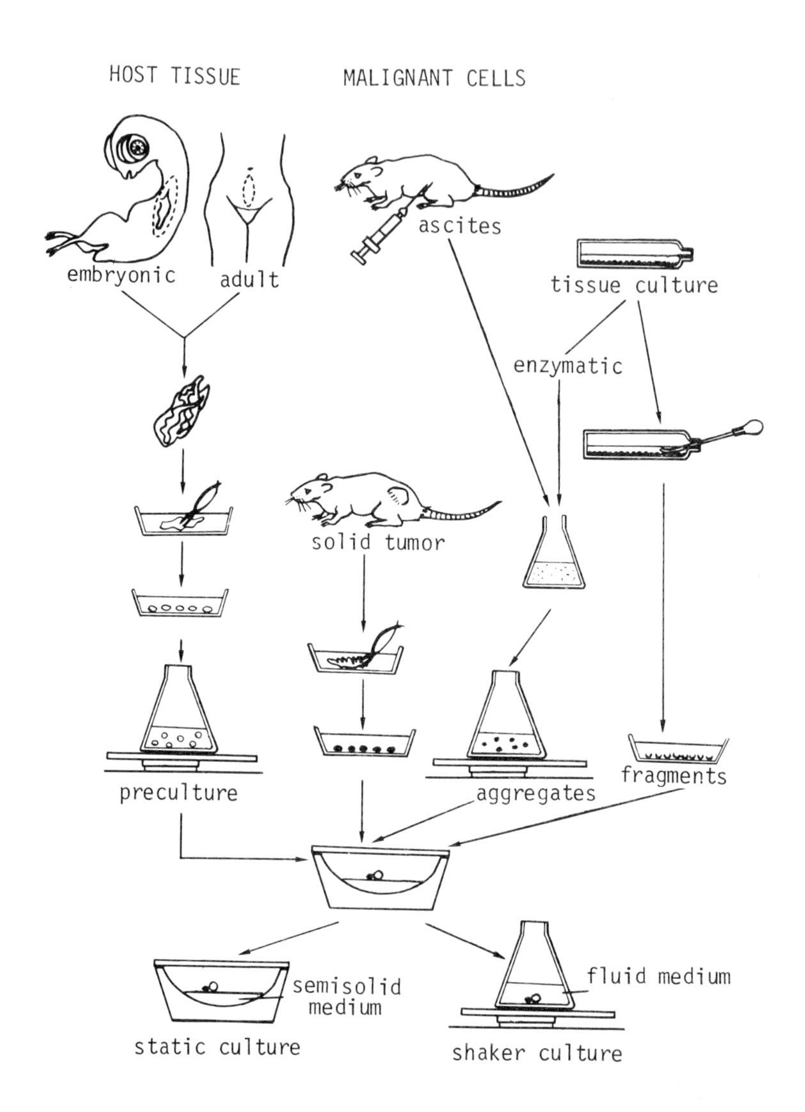

Fig. 1. General outline of the in vitro method for the study of invasion by malignant cells

The following points shall be considered.
1) Techniques of preparation and culture of the host tissue are designed to preserve its histotypical structure. To avoid central necrosis due to anoxia, the initial diameter of the fragment of host tissue is kept at about 0.4 mm.
2) The malignant cells are allowed to adhere to the host tissue by bringing both into physical contact. The time necessary to bring about adhesion (usually between 15 min and 4 h) depends on the types of the confronting tissues.
3) The use of substrates, to which cells do not adhere, limits cellular interaction to the confronting tissues. Interaction of the confronting tissues with the substrate complicates the analysis of invasion.

4) Invasion is followed in the course of time by fixation of individual cultures after various periods of incubation (usually 4 h, 1, 4, 7 and 14 days).
5) Entire confrontations are analyzed histologically by preparing serial paraffin sections. This analysis of the supracellular organization of the confronting tissues always preceeds ultrastructural observations of selected areas.
6) Control cultures are made either of host tissue alone or of host tissue confronted with non-malignant counterparts, if the latter are available.
7) Fragments of host tissue are precultured for 24 to 72 h to allow healing of the dissection trauma, so that the malignant cells are confronted initially with healthy host cells. In the case of a fragment of embryonic chick heart, preculturing results in the formation of a few layers of fibroblastic cells around the bulk of heart tissue, which consists mainly of muscle cells. These peripheral cells are usually not closely apposed to one another as e.g. endothelial cells are. They overlap each other and produce a fuzzy electrondense extracellular matrix, which sometimes organizes into fibers. Considering the structure of the periphery of precultured heart fragments, we can imagine that small extensions from confronting cells can penetrate between the fibroblastic cells without alteration of their disposition. On the other hand, considering that this peripheral layer functions as a natural lining of an isolated tissue fragment, we may presume that the disposition of the fibroblastic cells will be altered whenever the fragment of heart tissue fuses with another tissue, be it malignant or not (Fig. 2).

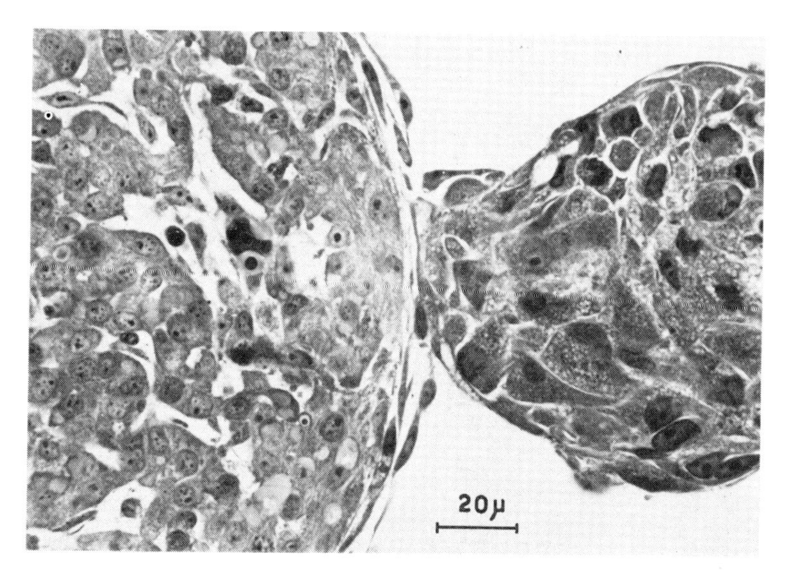

Fig. 2. Light micrograph of a 2 μm thick section from a confrontation of an aggregate of normal human fibroblasts (right) with a precultured heart fragment (left), fixed after 2 h incubation on a semi-solid medium. Hematoxylin and eosin.

8) Spheroidal aggregates of malignant cells are prefered to fragments of monolayer or single cells for confrontation with the host tissue in vitro. Their three-dimensional structure, their growth pattern, their metabolism and their reactions to anti-cancer agents resembles more closely that of a tumour in vivo than it is the case with e.g. a cell suspension or a monolayer. As an example, Fig. 3 shows the incorporation of ^{3}H-thymidine into DNA by MO_4 cells (malignant virally transformed C_3H mouse fibroblastic cells), grown as individual aggregates in shaker culture.

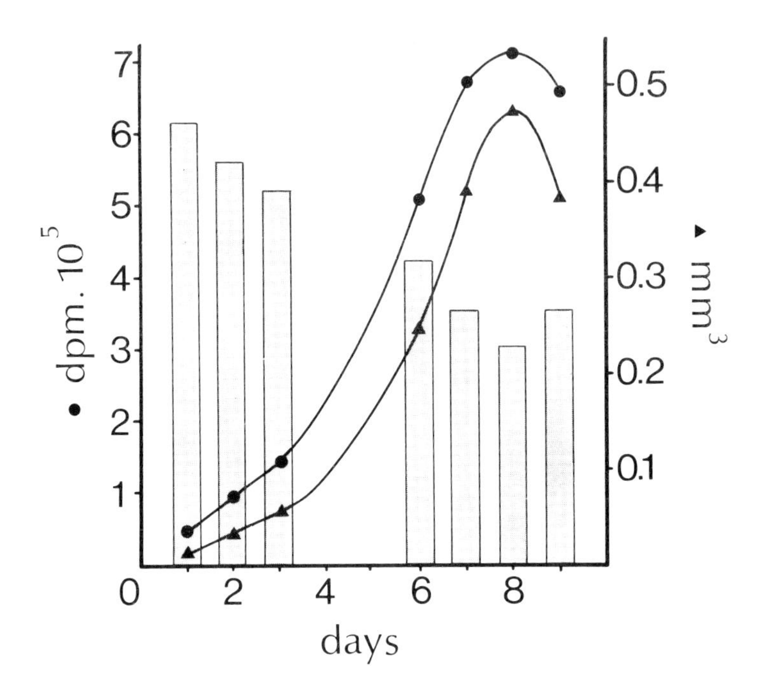

Fig. 3. Volume (right) and perchloric acid precipitable (left) ^{3}H-activity of individual aggregates of MO_4 cells, grown in shaker culture for 1 to 9 days. Bars indicate radioactivity per unit volume.

It is obvious that the increase in DNA synthesis does not parallel the increase in volume of the aggregates. This corresponds with the histological and autoradiographical data, which show that in the center of the aggregate cells stop dividing and become necrotic. Then, growth is due mainly to proliferation of cells in the peripheral layers. Such a growth pattern is also found in tumours in vivo (Aherne and co-workers, 1977).
Despite the advantages of three-dimensional aggregates of cells, prepared usually from suspensions of cells (Mareel, Kint and Meyvisch, 1979), we shall consider the possibility that they are formed by a selection of e.g. the most cohesive cells, which is not necessarily representative of the entire population.

INVASION IN VIVO AND IN VITRO

We have used the method described above to examine the invasiveness of various cell lines in vitro. A good correlation between invasiveness in vitro and the capacity to form invasive tumours in vivo is found in cell families that were tested before and after malignant alteration (Mareel and co-workers, 1975; Kieler and co-workers, 1979). These experiments strengthen the idea that invasiveness in vitro is a criterion of malignancy (for review see Mareel, 1979).
Nevertheless, the following differences between invasion in vitro and invasion in vivo shall be kept in mind.

1) In vivo, malignant cells arise inside the host and most of them pass through a non-invasive state (e.g. carcinoma in situ). In vitro, cells that have already acquired the capacity to invade are put next to freshly prepared host tissues. Therefore in vitro models are closer to secondary invasion, as it occurs in metastasis, than to primary invasion.
2) Malignant cells can be confronted in vitro with a variety of tissues, which do conserve their histotypical structure during culture. A few tissues, which are important for the spread of malignant cells in the body, cannot be used successfully as hosts in vitro. Amongst them are functional blood- and lymphvessels (intravasation) and the basement membrane (apical side).
3) Humoral and cellular reactions (e.g. inflammation) are absent during invasion in vitro.
4) In most experiments in vitro host tissues are derived from healthy individuals, whereas in vivo malignant cells invade into tissues, that belong to a tumour bearing individual.

CELLULAR ACTIVITIES DURING INVASION

Confrontations between malignant cells and host tissue in vitro are also used for the study of cellular activities, which are presumed to be involved in invasion. Growth and locomotion are discussed by G. Storme in this volume. We report here the value of the in vitro method for the study of : 1) Adhesion of malignant cells to the host tissue; 2) Destruction of the host tissue by invading malignant cells; 3) Phagocytosis of material from the host by the malignant cells.

Adhesion

Adhesion of the malignant cells to the host tissue appears to be the trigger of all other cellular activities involved in invasion in vitro (De Ridder and co-workers, 1975). In our assay (Fig. 4) an aggregate of malignant cells is considered to adhere to a precultured fragment of host tissue when both remain attached to one another when they move through the culture medium under the influence of gravity. Adhesion is quantitated through the number of adhering fragments per total number of confrontations tested in function of time.
In this assay, the adhesion of aggregates of MO_4 cells to precultured fragments of embryonic chick heart, is inhibited by low temperature (4° C) and by cytochalasin B, but not by the microtubule inhibitor nocodazole or by 5-fluorouracil. Transmission electron micrographs from MO_4 aggregates adhering to heart fragments and fixed 30 min after the initial contact show long multifarious extensions from MO_4 cells penetrating between the peripheral fibroblastic cells of the heart fragment. These extensions frequently end in filopodia, 0.1 µm in diameter, which contain microfilaments and are similar to the extensions of invading cancer cells in vivo (Gabbiani, Trenchev and Holborow, 1975).

Destruction

In the in vitro assay of invasiveness, destruction of the host tissue by the invading malignant cells is suspected from 1) necrotic material visible histologically or ultrastructurally at the front of invasion and 2) progressive replacement of the host tissue by malignant cells. In experiments where MO_4 cells are allowed to invade from an aggregate into a fragment of embryonic chick heart, total replacement of the heart can be demonstrated histologically in serially fixed cultures. This observation is confirmed by comparison of the lactate dehydrogenase isozyme pattern of confrontations, that are examined different times after the initial contact between the MO_4 cells and the heart (Fig. 5).

The mechanism of destruction of the host tissue by invading malignant cells remains unknown.

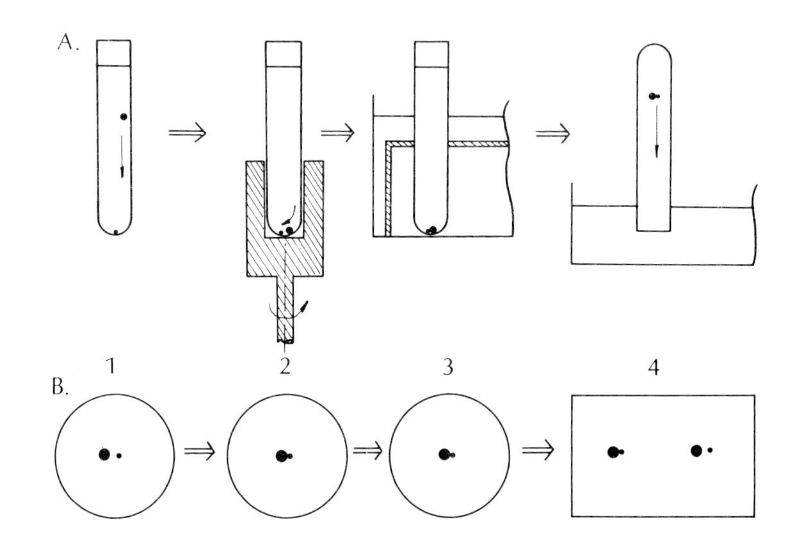

Fig. 4. Outline of the assay of adhesion. A fragment of pre-cultured host tissue and a spheroidal aggregate are put in a centrifuge tube (1) and brought into physical contact by rotation at 50 rpm. After incubation (3) the tube is turned upside down (4). Adhesion (left) and absence of adhesion (right) are scored under the stereomicroscope (4, bottom).

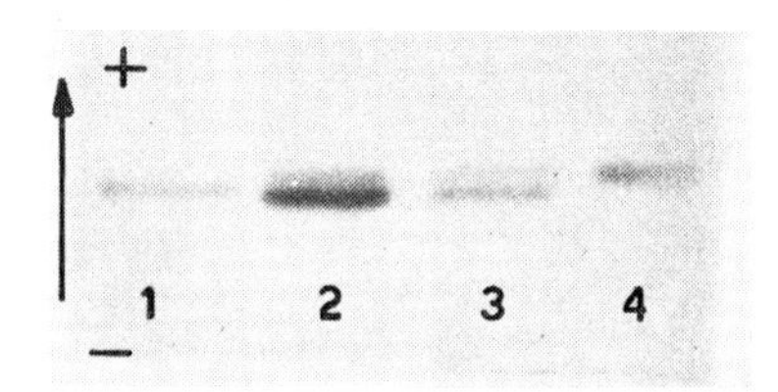

Fig. 5. Cellulose-acetate electrophorogram of homogenates from precultured heart fragments alone (1), and confronted with an aggregate of MO_4 cells for one (2), three (3) and seven (4) days in shaker culture.

Phagocytosis

Attempts to reveal destruction of the host tissue by measurement of ^{51}Cr-release into the culture medium failed (Mareel, De Bruyne and De Ridder, 1977). One possible explanation of this failure is phagocytosis of material from the host tissue by the malignant cells.

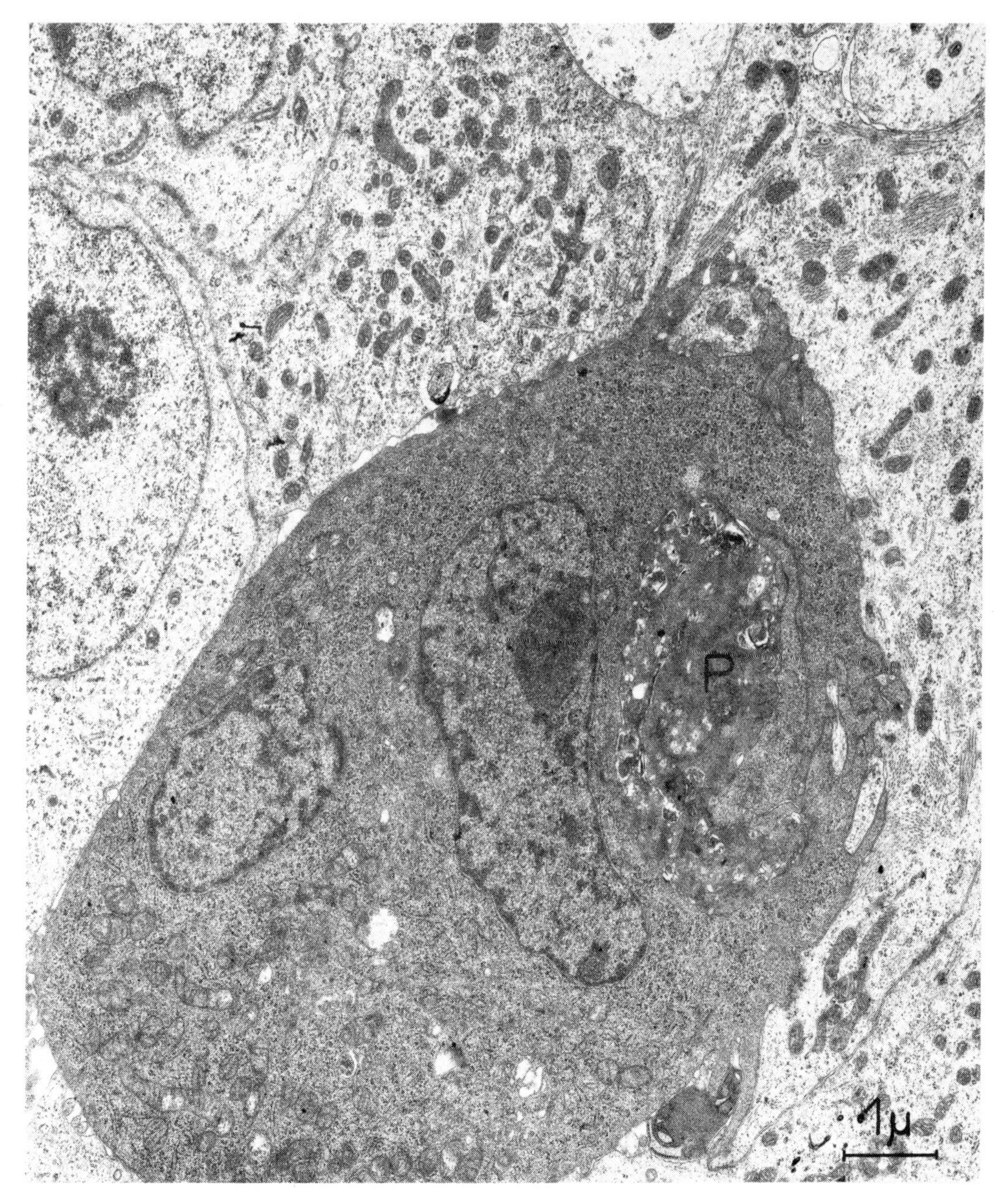

Fig. 6. Transmission electronmicrograph of a HeLa cell invading into embryonic chick heart tissue in vitro. P : phagosome.

Experiments with Suramin (Gift from Bayer) make further investigations attractive, since they indicate that phagocytosis might be an important part of invasion. This lysosomotropic drug inhibits invasion in vitro, and we explain this inhibition by the fact that the drug interferes with the fusion of lysosomes and phagosomes (D'Arcy Hart and Young, 1975). Distinction between autophagosomes, which occur frequently in malignant cells, and heterophagosomes is one of the major problems in studying the role of phagocytosis during invasion (Fig. 6).
In collaboration with F. Van De Sande (Department of Embryology, State University Ghent) we have started to solve this problem, taking advantage of the heterotypic character of the in vitro confrontation of cells derived from different species. Using an anti-serum against chick heart, prepared in rabbits, fragments of heart cells are demonstrated inside vacuoles of invading MO_4 cells or HeLa cells by the PAP-immunocytochemical method (Van De Sande, 1979).

CONCLUSION

Confrontation of malignant cells with precultured tissue fragments in three-dimensional culture constitutes a valuable method to study invasion, provided differences between invasion in vivo and in vitro are kept in mind. The in vitro method can be used for the examination of the invasiveness of various cell types and for the analysis of cellular activities involved in invasion. The method allows histological, ultrastructural, metabolic and biochemical observations.

ACKNOWLEDGMENT

This work is supported by a Grant from the Kankerfonds van de Algemene Spaar- en Lijfrentekas, Brussels, Belgium.

REFERENCES

Aherne, W.A., R.S. Camplejohn, M. Al-Wiswasy, D. Ford, and A.M. Kellerer (1977). Assessment of inherent fluctuations of mitotic and labeling indices of human tumours. Brit. J. Cancer, 36, 577-582.

D'Arcy Hart, P., and M.R. Young (1975). Interference with normal phagosome-lysosome fusion in macrophages, using ingested yeast cells and suramin. Nature, 256, 47-49.

De Ridder, L., M. Mareel, and L. Vakaet (1975). Adhesion of malignant and non-malignant cells to cultured embryonic substrates. Cancer Res., 35, 3164-3171.

Gabbiani, G., P. Trenchev, and E.J. Holborow (1975). Increase of contractile proteins in human cancer cells. Lancet, II, 796-797.

Kieler, J., P. Briand, M.C. Van Peteghem, and M. Mareel (1979). Comparative studies of two types of "spontaneous" malignant alteration of ST/a mouse lung fibroblasts propagated in vitro. In Vitro, in Press.

Mareel, M. (1979). Is invasiveness in vitro characteristic of malignant cells ? Cell Biology International Reports, in Press.

Mareel, M., L. De Ridder, M. De Brabander, and L. Vakaet (1975). Characterization of spontaneous, chemical, and viral transformants of a CH3/3T3-type mouse cell line through transplantation into young chick blastoderms. J. Natl. Cancer Inst., 54, 923-929.

Mareel, M., G. De Bruyne, and L. De Ridder (1977). Invasion of malignant cells into ^{51}Cr-labeled host tissues in organotypical culture. Oncology, 34, 6-9.

Mareel, M., J. Kint, and C. Meyvisch (1979). Methods of study of the invasion of malignant C_3H-mouse fibroblasts into embryonic chick heart in vitro. Virchows Arch. B Cell Path., 30, 95-111.

Meyvisch, C., and M. Mareel (1979). Invasion of malignant C_3H mouse fibroblasts from aggregates transplanted into the auricles of syngenic mice. Virchows Arch. B Cell Path., 30, 113-122.

Van De Sande, F. (1979). A critical review of immunocytochemical methods for light microscopy. J. Neuroscience Methods, 1, 3-23.

Quantitative Analysis of Invasiveness *in vitro*

L. de Ridder

Department of Experimental Cancerology, Clinic of Radiotherapy and Nuclear Medicine, State University, Academic Hospital, B-9000 Ghent, Belgium

ABSTRACT

Confrontations of functional whole thyroid lobes from 12 to 13 days old embryonic chicks with invasive MO_4 cells are cultured in a three-dimensional system in vitro. Comparison by histology, autohistoradiography and gammacounting of I^{125} uptake of control thyroid lobes and confrontations, reveals differences in the morphology and function of both groups. Irreversible structural alterations and a pronounced decrease in inorganic iodine uptake in the confrontations can be correlated with the extent of invasion in the thyroid. Measurements of the iodine uptake in confrontations may serve as a method for quantitative evaluation of invasion.

KEYWORDS

Invasion; quantitation; in vitro; embryonic thyroids; iodine uptake; functional morphology.

INTRODUCTION

Extension of a malignant tumour occurs through growth and invasion and is usually accompanied by destruction and replacement of host tissues. The histology of many malignant tumours suggests that solitary cells and strands of cells leave the main mass and invade into the neighbouring tissues. This is in striking contrast with most benign tumours, so that invasiveness is considered as an essential feature of malignant cells. It is tempting to speculate that the degree of malignancy of a given tumour correlates with the invasive potential of the composing malignant cells.

Obviously, quantitative evaluation of both malignancy and invasive potential are needed to test this hypothesis. For the evaluation of malignancy several techniques are available at least for animal tumours, but a reliable and generally applicable method to measure invasiveness is missing.

Attempts to evaluate the invasive potential of tumour cells from the histological aspect of biopsy specimens have met with great difficulties. E.g., the depth of invasion tells nothing about the invasive capacity of the tumour cells unless we know the moment the first malignant cell started to invade. This is usually not the case with spontaneous or transplanted tumours in vivo (Meyvisch and Mareel, 1979).

Confrontation of malignant cells with fragments of normal tissues in three-dimensional culture (Easty and Easty, 1963; Schleich, Frick and Mayer, 1976; De Ridder, Mareel and Vakaet, 1977; Mareel, Kint and Meyvisch, 1979) mimics invasion of malignant tumours. There exists a good correlation between the invasiveness of cells in these in vitro systems and their capacity to form invasive tumours in vivo (for review see Mareel, 1979).
Using the chick chorio-allantoic membrane as a host tissue, Easty and Easty (1974) and Hart and Fidler (1978) have quantitated invasion by measuring the depth of penetration and by counting the number of cells passing through the membrane respectively. In both methods the activity of the invasive cells is considered directly.
We have tried to infer the activity of the invading malignant cells, indirectly, from alterations of the host tissue.
Chick embryonic thyroids are used as host tissue, because alterations of their structure and function caused by invading malignant cells can be evaluated both morphologically and metabolically.

MATERIALS AND METHODS

The thyroid lobes are dissected from 12 to 13 days old chick embryos. At this age the thyroid is functional in ovo as is evident from the selective uptake of iodine and from the production of hormones (Daugeras and co-workers, 1975) and in vitro (De Ridder and Mareel, 1978).
Entire thyroid lobes are confronted with malignant mouse fibroblastic cells (MO_4) in gyratory shaker culture, as described by Mareel in this volume, using an atmosphere of 5% CO_2 in air. In all experiments control thyroids are cultured without MO_4 cells. The malignant cells are confronted with the thyroid in one of the following ways. 1) Thyroids are cultured in a medium containing $10^5 MO_4$ cells /ml harvested from monolayer cultures by trypsinisation. 2) A fragment from a monolayer is put on top of a thyroid, allowed to adhere for 2 h and transferred to the shaker culture. 3) An aggregate of MO_4 cells is put next to a thyroid (Fig. 1) and processed as described for 2). Cultures are run for 1 to 4 days and Na I^{125} is added to the culture medium 24 h before termination.
Tissues are washed in cold culture medium, fixed in glutaraldehyde and their radioactivity is measured with a gammacounter. The uptake of iodine by the thyroid is expressed as the number of counts/min per mm^3 of thyroid divided by the number of counts/min per mm^3 of medium. The volume of the thyroid is measured at the onset of the culture. After gammacounting the tissues are processed for histology. Alternative 2 µm thick sections are stained for light microscopy or coated with a liquid emulsion (Ilford L_4) and exposed for histoautoradiography.

RESULTS

Histology

Individual control thyroid lobes are surrounded by a capsule and a small layer of connective tissue, which extends into the lobe separating individual follicles. In vitro culture causes distension of follicles, which is most pronounced in the center. Central necrosis invariably occurs when the diameter of the lobe is greater than 0.6 mm.
Invasion of MO_4 cells is demonstrated in the three types of confrontation.

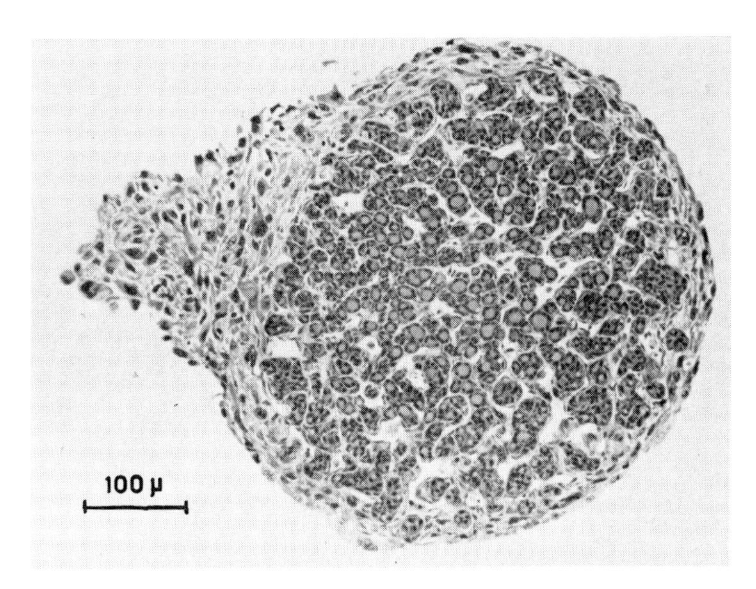

Fig. 1. Light micrograph of a 2 µm thick section from a confrontation of a fragment from a MO_4 monolayer with a thyroid lobe, fixed after 24 h in shaker culture. Hematoxylin and eosin.

Strands and solitary MO_4 cells are found inside the thyroid tissue. These invading MO_4 cells are mostly spindle shaped, whereas, in the confronting bulk, MO_4 cells are mostly polygonal. Vacuoles are frequently seen inside invading cells and presumably reveal their phagocytic activity.
The histology of cultures fixed after various periods of incubation shows progressive replacement of thyroid tissues by MO_4 cells. Complete replacement of the whole lobe is not observed. In most cultures it is difficult to distinguish between central necrosis due to anoxia and necrosis at the front of invasion. The capsule and the connective tissue are usually replaced earlier than the follicles, so that follicles surrounded by MO_4 cells closely apposed to their basal side are present in many cultures. Disintegration, partial and complete replacement of follicles can be demonstrated in our histological material beyond any doubt.

Histoautoradiography

No label is found over MO_4 cells, cultured alone or with thyroid tissue in presence of radioactive iodine.
In control thyroids and in thyroids confronted with MO_4 cells the intensity of the label decreases from the periphery towards the centre. Label is always concentrated over the follicles, although large individual variations in labeling intensity are obvious. As a rule, a morphologically intact follicle, be it distended or completely surrounded by invading MO_4 cells, appears to be able of concentrating radioactive inorganic iodine into its lumen (Fig. 2B). Contrary, cells from disintegrated follicles do not concentrate the isotope.

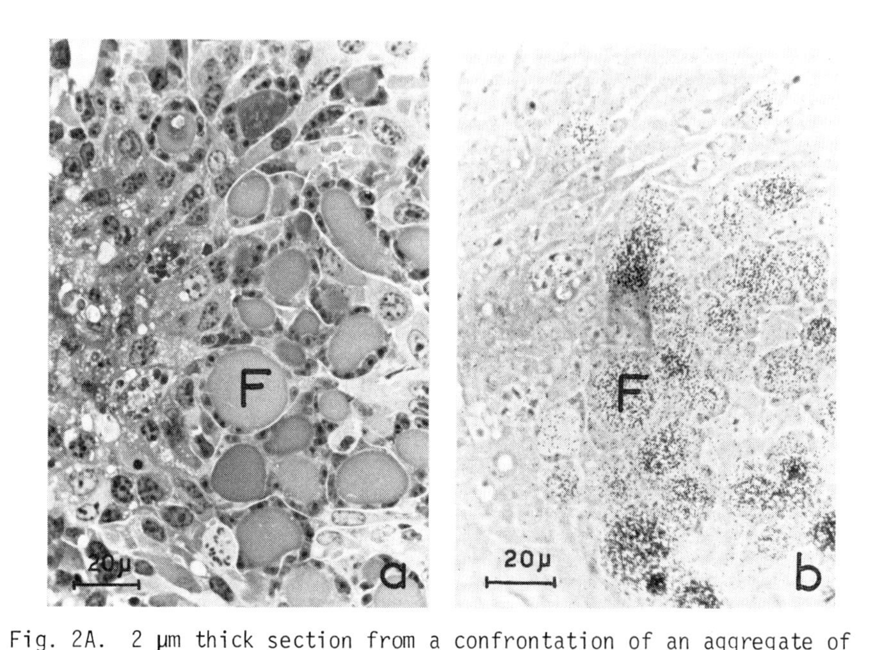

Fig. 2A. 2 µm thick section from a confrontation of an aggregate of MO_4 cells (left) with a thyroid lobe (right) fixed after 3 days. F = follicle. Hematoxylin and eosin.
2B. Histoautoradiogram of a section following that presented in Fig. 2A. Note concentration of label over distended follicles.

Our histoautoradiograms show that the uptake of inorganic iodine depends on the number of intact follicles and on the amount of iodine concentrated by the individual follicles. Although it is readily seen from serial sections that invasion by MO_4 cells diminishes the capacity of the thyroid to incorporate iodine, histoautoradiographs are not suitable for quantitation of invasion.

Gammacounting

The iodine uptake in controls is compared with the uptake in thyroids confronted with aggregates of MO_4 cells. This type of confrontation is choosen because the volume of the aggregate allows an estimation of the number of MO_4 cells, a factor that is supposed to influence invasion quantitatively (Mareel, Kint and Meyvisch, 1979).
Large variations in the uptake of radioactive iodine in control thyroids is the main drawback of our study model. It is not yet clear to us if these variations have to be ascribed to variations in the function of individual thyroids or to pitfalls in culture conditions. Loss of function during culture of controls constitutes an additional difficulty.
Nevertheless, within the same series of experiments loss of function is always more pronounced in thyroids that are invaded by MO_4 cells (Fig. 3).

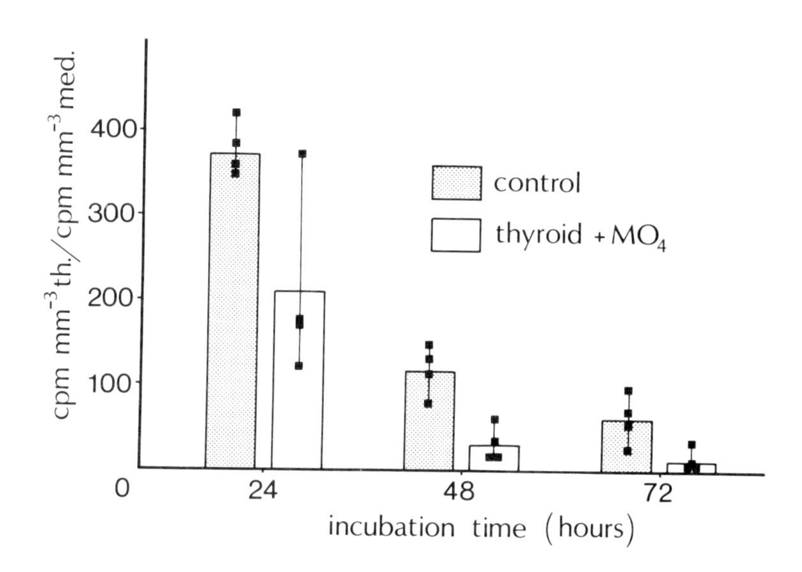

Fig. 3. Diagram showing the number of counts per minute per mm^3 thyroid divided by the number of counts per minute per mm^3 medium after different incubation times.

The iodine uptake in controls and in confronting cultures is expressed as the radioactivity present in the tissue at the end of the culture period divided by the volume of the thyroid at the onset of the experiment.
Calculations of the volumes from measurements of the diameter on stereomicrographs made after various periods of incubation show a decrease of the volume in controls and an increase in confronting tissues.
The latter, obviously, is due to growth of the MO_4 cells. Since confronting MO_4 cells not only invade into the thyroid, but also have a tendency to surround it, we cannot exclude that the central anoxic area of the thyroid is larger in confronting cultures than in controls. If this is the case, loss of function not directly due to invasion has to be considered. Experiments with non-malignant cells, which do surround the thyroid without invading it, indicate that this loss of function due to anoxia in the centre of the culture, cannot account for the differences between controls and invaded thyroids as shown in Fig. 3.

CONCLUSIONS

Confrontations of thyroid lobes with malignant MO_4 cells in three-dimensional culture, demonstrate morphological and functional alterations of the thyroid tissues due to invading MO_4 cells. Measurements of the uptake of inorganic iodine in controls and in invaded thyroids support the idea that invasion can be evaluated indirectly through alterations of the host tissue.
Before the present study model, however, can be used for quantitation of invasion several problems remain to be solved. Central necrosis, which is responsible for

the loss of function in controls, can be prevented by the use of an atmosphere containing a high concentration of oxygen (De Ridder and Mareel, 1978). Such a concentration of oxygen, however, might be toxic for malignant cells (Pace, Thompson and Van Camp, 1962; Richter, Sanford and Evans, 1972). Since central anoxia depends on the volume of the culture tissue, fragments of thyroid might be worth trying. It is however not excluded that fragmentation of thyroid lobes may lead to loss of differentiation and thus to a pronounced decrease of function.
Large variations in the function of controls might be due to biological variations that are typical for the thyroid. This factor counteracts the major advantage of thyroid tissue, namely its capacity to concentrate inorganic iodine in a highly specific way.
We are actually exploring the possibility to measure the function of other tissues, that might serve as a host tissue for the invasion of malignant cells in vitro.

REFERENCES

Daugeras, M., A. Brissan, F. Lapointe-Boulou, and F. Lachiver (1976). Thyroidal iodine metabolism during the development of the chick embryo. Endocrinology, 95, 442-458.
De Ridder, L., M. Mareel, and L. Vakaet (1977). Invasion of malignant cells into cultured embryonic substrates. Arch. Geschwulstforsch., 47, 7-27.
De Ridder, L., and M. Mareel (1978). Morphology and ^{125}I-concentration of embryonic chick thyroids cultured in an atmosphere of oxygen. Cell Biology International Reports, 2, 189-194.
Easty, D., and G. Easty (1974). Measurement of the ability of cells to infiltrate normal tissues in vitro. Brit. J. Cancer, 29, 35-49.
Easty, G., and D. Easty (1963). An organ culture system for the examination of tumour invasion. Nature (London), 199, 1104-1105.
Hart, I., and J. Fidler (1978). An in vitro quantitative assay for tumor cell invasion. Cancer Res., 38, 3218-3225.
Mareel, M., J. Kint, and C. Meyvisch (1979). Methods of study of the invasion of malignant C_3H-mouse fibroblasts into embryonic chick heart in vitro. Virch. Arch. B Cell Path., 30, 95-111.
Mareel, M. (1979). Is invasiveness in vitro characteristic of malignant cells ? Cell Biology International Reports, in Press.
Meyvisch, C., and M. Mareel (1979). Invasion of malignant C_3H mouse fibroblasts from aggregates transplanted into the auricles of syngenic mice. Virch. Arch. B Cell Path., 30, 113-122.
Pace, D., J. Thompson, and W. Van Camp (1962). Effects of oxygen on growth in several established cell lines. J. Natl. Cancer Inst., 28, 897-909.
Richter, A., K. Sanford, and V. Evans (1972). Influence of oxygen and culture media on plating efficiency of some mammalian tissue cells. J. Natl. Cancer Inst., 49, 1705-1712.
Schleich, A., M. Frick, and A. Mayer (1976). Patterns of invasive growth in vitro. Human decidua graviditatis confronted with established human cell lines and primary human explants. J. Natl. Cancer Inst., 56, 221-237.

Influence of Anti-cancer Agents on Growth, Migration and Invasion of Malignant Fibroblastic Cells

G. Storme and M. Mareel*

Oncology Center, Radioth., Free University Brussels, Acad. Hosp., Laarbeeklaan 101, B-1090, Belgium

**Dept. of Exp. Cancerology, Clin. of Radioth. and Nucl. Med., Acad. Hosp., B-9000 Ghent, Belgium*

ABSTRACT

The effect of 5-fluorouracil, ionizing radiation and the microtubular inhibitor nocodazole on growth and directional migration of virally transformed malignant C3H mouse fibroblastic cells (MO_4) is compared with their effect on the invasiveness of these cells in vitro. The increase in diameter of individual aggregates of MO_4 cells in shaker culture is used as an index of growth. The mean diameter of the circular area covered by MO_4 cells migrating from an aggregate explanted on glass is used as an index of directional migration. Invasion is studied using confrontations of aggregates of MO_4 cells with precultured fragments of embryonic chick heart in shaker culture. Ionizing radiation (5000 R) and 5-fluorouracil (1 µg/ml) permit directional migration and invasion for at least 14 days after the onset of treatment, although they inhibit growth. The microtubule-inhibitor nocodazole (1 µg/ml) inhibits growth and prevents both directional migration and invasion. We conclude that various anti-cancer agents, at doses that inhibit growth, have different effects on invasion. We suggest that the assay of directional migration is used for the screening of potential anti-invasive agents.

KEYWORDS

Invasion; growth; directional migration; ionizing radiation; 5-fluorouracil; microtubule-inhibitors; malignant cells; cellular aggregates.

INTRODUCTION

Invasion into the surrounding tissues, contributes to the malignant character of most tumours. To our knowledge this primary invasion has not been considered as a target for therapy of malignant tumours. We report here the influence of 5-fluorouracil (Roche Laboratories, Nutley, N.J.), nocodazole (Janssen Pharmaceutica, Beerse, Belgium) and ionizing radiation on invasion, growth and migration of virally transformed malignant C3H mouse fibroblastic cells (MO_4) (Billiau and coworkers, 1973). MO_4 cells are maintained in tissue culture on glass or plastic. Suspensions of MO_4 cells, harvested from tissue culture by trypsinization, produce spheroidal aggregates, the growth of which can be followed via the increase of their diameter. These aggregates produce invasive fibrosarcomas when transplanted into syngenic mice (Meyvisch and Mareel, 1979).

INVASION IN VITRO

Mareel and De Brabander (1978) have shown that microtubule-inhibitors prevent invasion of MO_4 cells into embryonic chick heart in organotypic culture on semi-solid medium, contrary to a series of other growth-inhibitors.

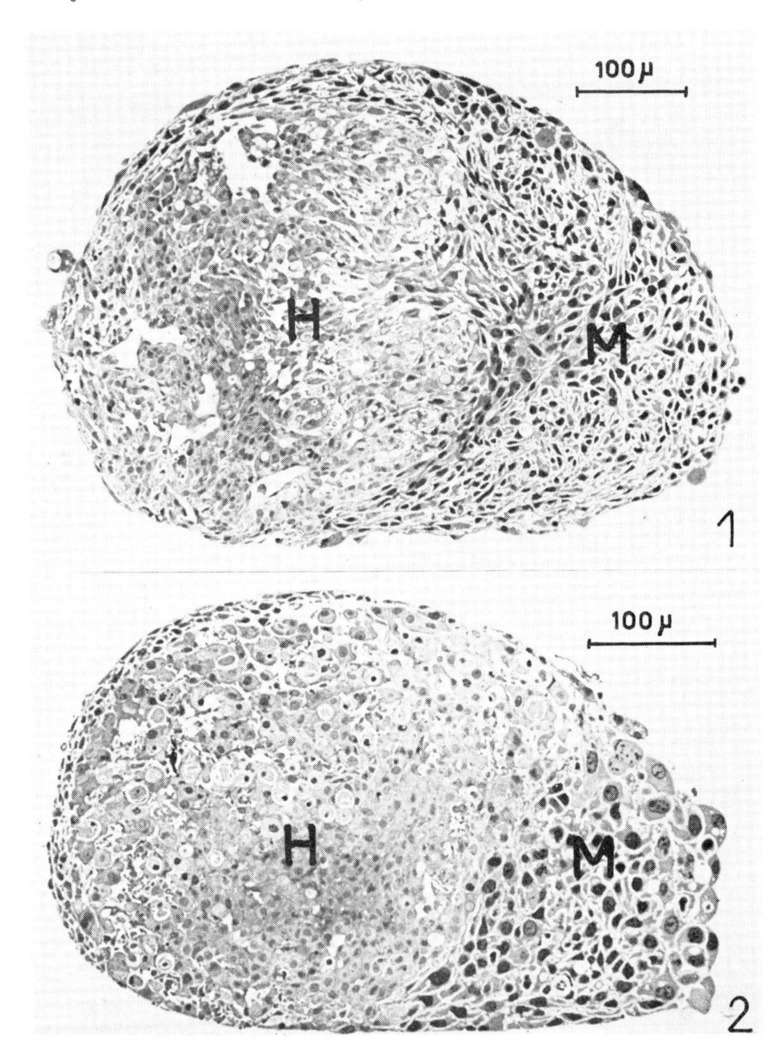

Fig. 1 and 2. 2 µm thick sections from confrontations of MO_4 cells (M) and embryonic chick heart (H) in presence of 1 µg/ml (1) and 10 µg/ml (2) R 45911 (Janssen Pharmaceutica, Beerse, Belgium). Fixation after 4 days. Partial (1) and complete (2) inhibition of invasion; in control cultures the heart tissue is almost completely replaced by MO_4 cells after 4 days. Hematoxylin and eosin.

We have repeated this experiment in shaker culture : 5-fluorouracil (1 µg/ml) and ionizing radiation (5000 R) inhibit growth but permit invasion, whereas the microtubular-inhibitor nocodazole (1 µg/ml) inhibits invasion (Storme and Mareel, 1979). It appears to us that confrontation of spheroidal aggregates of MO_4 cells with precultured fragments of embryonic chick heart in shaker culture (Mareel, Kint and Meyvisch, 1979) is a useful model for testing the anti-invasive effect of new microtubule-inhibitors (Figs. 1 and 2).

GROWTH IN SHAKER CULTURE

To study the effect of anti-cancer agents on growth, aggregates of MO_4 cells, with a diameter of about 0.2 mm, are cultured individually in 5 ml Erlenmeyer flasks containing 1.5 ml of medium on a gyratory shaker at 120 rpm. Growth is evaluated by measuring the diameter of the spheroidal aggregates at regular intervals (Figs. 3 and 4).

In spheroidal aggregates of cells the increase of the diameter of the horizontal projection is an estimation of growth which mimics tumour growth in vivo (Sutherland, McCredie and Inch, 1971; Yuhas and co-workers, 1977). This estimation of growth takes into account various factors which determine the biological behaviour of the whole cell population. The growth of the aggregate is the result of both, cell proliferation and cell loss (Durand, 1975; Yuhas and Li, 1978). In spheroidal cell aggregates in vitro and in tumours in vivo cell loss is caused by exfoliation and by entrance of cells into the central necrotic area. The latter, which is a consequence of growth without neovascularisation, has an influence on the survival and the dissemination of the malignant cells. Additional factors influencing growth are individual cell volume and intercellular space. The effect of ionizing radiation, 5-fluorouracil, nocodazole and betamethasone on the growth of aggregates of MO_4 cells is shown in fig. 5.

Histologic examination of paraffin sections from treated aggregates indicates that inhibition of growth is due mainly to inhibition of cell division. Stereomicroscopy of several treated aggregates shows an irregular surface (Fig. 4) as compared to controls, the surface of which is usually smooth. Such irregular surfaces suggest shedding of cells into the culture medium. Further experiments are needed to determine cell loss in control and treated aggregates. We conclude from the present experiments that treatment with 5-fluorouracil, ionizing radiation or nocodazole can produce non-proliferating aggregates of MO_4 cells.

MIGRATION ON GLASS

MO_4 cells survive the block of cell division and are able to perform locomotion amongst other cellular functions. A spheroidal aggregate of MO_4 cells explanted on glass provides us with a simple assay of directional cell migration as defined by De Brabander and co-workers (1976). Quantitative analysis is done by measurement of the mean diameter of the circular area covered by the cells which migrate from the aggregate (Fig. 6).

Cinephotomicroscopy shows that cells, leaving the aggregate, reach the outer border of this area by active movement. Floating of cells and reattachment to the substrate was not observed. Why cells move away from the aggregate is not clear. This phenomenon seems to be triggered by contact with the artificial substrate and is not specific of malignant cells. The cinephotomicrographic and the histologic pattern of the cultures suggest that the migration is directional. The effect of 5-fluorouracil, ionizing radiation and nocodazole is shown in Fig. 7.

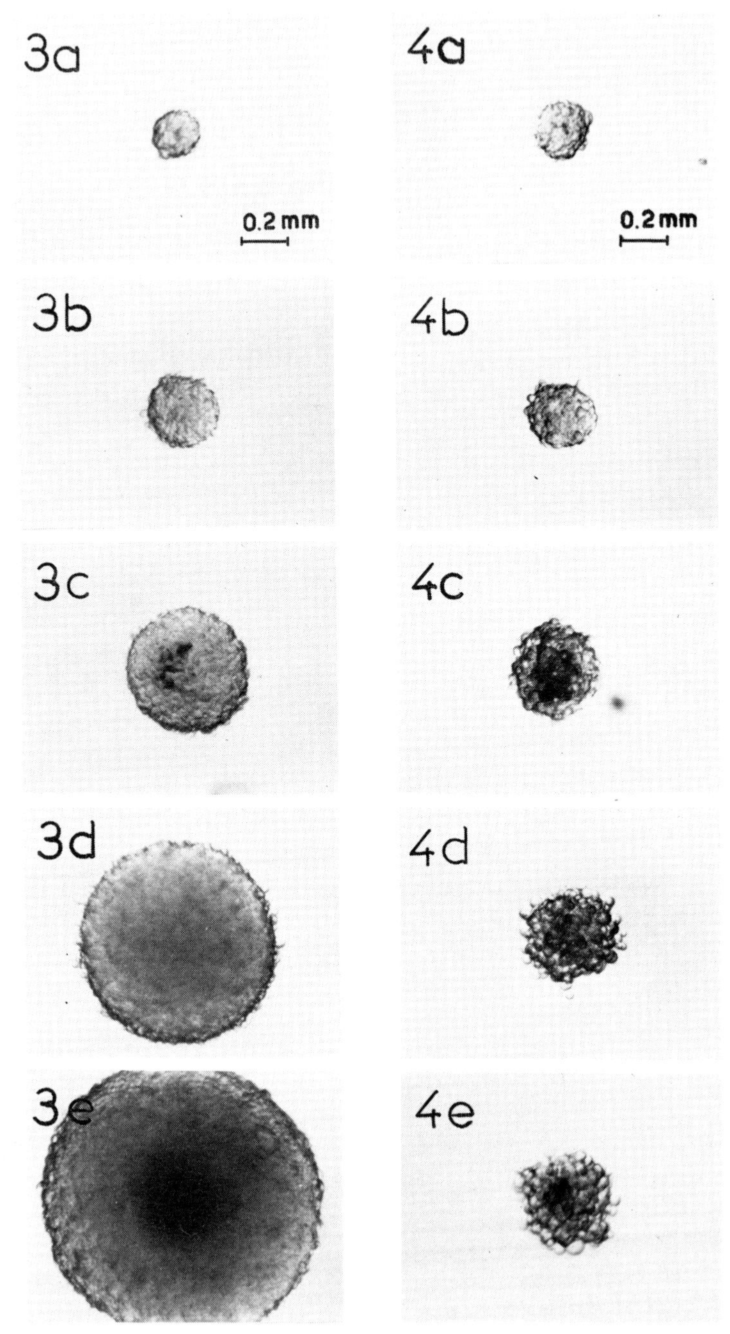

Figs. 3 and 4. Stereomicrographs of a control aggregate (3) and an aggregate irradiated with 5000 R (4), at the onset of the shaker culture (a), after 1 day (b), 3 days (c), 5 days (d) and 7 days (e).

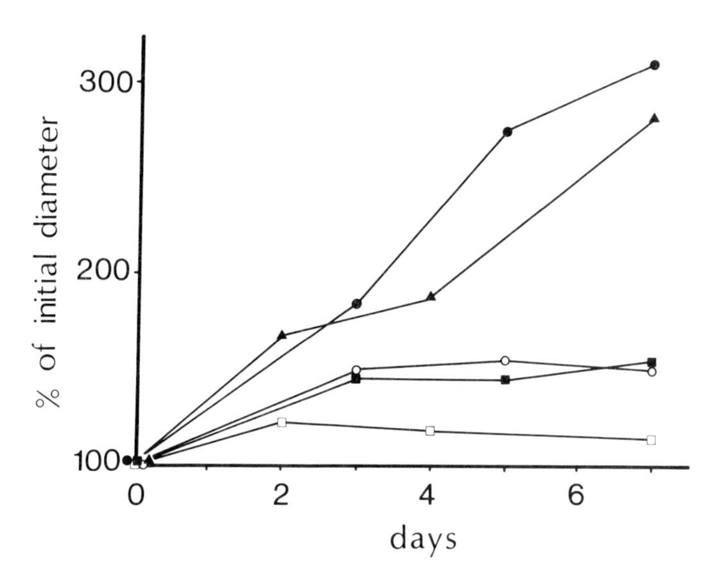

Fig. 5. Growth of spheroidal aggregates of MO_4 cells. Ordinate : % of original diameter (100% is about 0.2 mm); abscissa : time in days. Median values of 5 measurements. ● controls; ■ ionizing radiation (5000 R); ○ 5-fluorouracil (1 µg/ml); □ nocodazole (1 µg/ml); ▲ betamethasone (10 µg/ml).

Comparison of Fig. 5 with Fig. 7 indicates that growth and directional migration are basically unrelated phenomena.
Nevertheless, 5-fluorouracil and ionizing irradiation seem to have some inhibitory effect on directional migration as compared to controls, which becomes obvious after 4 to 7 days (Fig. 7). It is not excluded that this inhibitory effect is due to the effect of the agent on the number of cells rather than on the capacity of individual cells to migrate. Preliminary experiments have shown that the mean diameter of the area covered by MO_4 cells migrating from a treated aggregate explanted on glass, is to some extent related to the diameter of the original aggregate.
We have also examined how long treated MO_4 cells conserve their capacity to migrate on glass. Parallelly we have tested how long they conserve their ability to invade into fragments of embryonic chick heart in shaker culture. Aggregates of MO_4 cells are left in shaker culture for periods of 6 to 14 days after the onset of treatment. Then, they are either explanted on glass or confronted with fragments of embryonic chick heart during 4 days. Although the assay for directional migration reveals some inhibitory effect of 5-fluorouracil (1 µg/ml) and of ionizing radiation (5000 R), the latter experiments show that with both treatments the capacity to migrate on glass and to invade into the chick heart is conserved for at least 14 days. Quantitative comparison of migration and invasion cannot be done, because a quantitative assay for invasion is not as yet available.

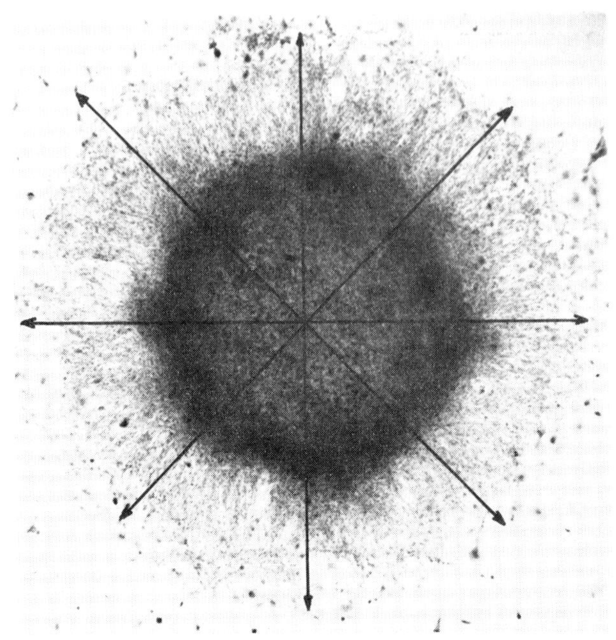

Fig. 6. Measurements of the circular area covered by the MO_4 cells migrating from the aggregate explanted on glass are indicated by the arrows. The mean value of the 4 measurements are plotted in function of time (see Fig. 7).

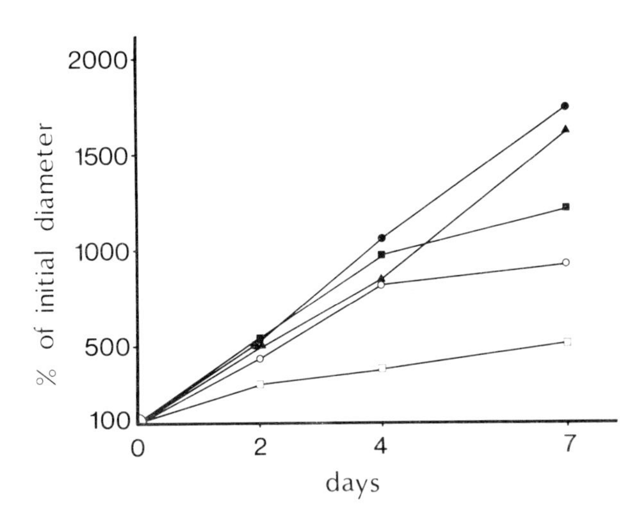

Fig. 7. Migration of MO_4 cells from an aggregate explanted on glass. Ordinate : % of initial diameter (100% is about 0.2 mm); abscissa : time in days. ● controls; ■ ionizing radiation (5000 R); ○ 5-fluorouracil (1 µg/ml); □ nocodazole (1 µg/ml); ▲ betamethasone (10 µg/ml). Median values from at least 5 cultures.

CONCLUSION

We conclude that anti-cancer agents, like 5-fluorouracil and ionizing radiation at doses, that inhibit growth, permit both directional migration and invasion. Microtubule-inhibitors, like nocodazole, at mitostatic doses, interfere with both directional migration and invasion. We therefore suggest that the assay of directional migration is used for the screening of large numbers of potential anti-invasive agents. Those agents, which are shown to inhibit directional migration, can then be tested in the in vitro assay for invasiveness.

REFERENCES

Billiau, A., H. Sobis, H. Eyssen, and H. Vanden Berghe (1973). Non-infectious intracysternal A-type particles in a sarcoma-positive leukemia-negative mouse cell line transformed by murine sarcoma virus (MSV). Arch. Ges. Virusforsch., 43, 345-351.

Durand, R.E. (1975). Isolation of cell subpopulations from in vitro tumor models according to cell sedimentation velocity. Cancer Res., 35, 1295-1300.

De Brabander, M.J., R. Van de Veire, F. Aerts, M. Borgers, and P. Janssen (1976). The effects of Methyl 5- (2-thienylcarbonyl)-1H-benzimidazole-2yl carbamate (R17934, NSC 238159), a new synthetic antitumoral drug interfering with microtubules on mammalian cells cultured in vitro. Cancer Res., 36, 905-916.

Mareel, M., and M. De Brabander (1978). Effect of microtubule inhibitors on malignant invasion in vitro. J. Natl. Cancer Inst., 61, 787-792.

Mareel, M., J. Kint, and C. Meyvisch (1979). Methods of study of the invasion of malignant C_3H-mouse fibroblasts into embryonic chick heart in vitro. Virch. Arch. B Cell Path., 30, 95-111.

Meyvisch, C., and M. Mareel (1979). Invasion of malignant C_3H-mouse fibroblast from aggregates transplanted into the auricles of syngenic mice. Virch. Arch. B Cell Path., 30, 113-122.

Storme, G., and M. Mareel (1979). Invasiveness of non-proliferating malignant cell populations. Europ. J. Cancer. In Press.

Sutherland, R.M., J.A. McCredie, and R.W. Inch (1971). Growth of multicell spheroids in tissue culture as a model of nodular carcinomas. J. Natl. Cancer Inst., 46, 113-120.

Yuhas, J.M., A.P. Li, A.O. Martinez, and A.J. Ladman (1977). A simplified method for production and growth of multicellular tumor spheroids. Cancer Res., 37, 3639-3643.

Yuhas, J.M., and A.P. Li (1978). Growth fraction as the major determinant of multicellular tumor spheroids growth rates. Cancer Res., 38, 1528-1532.

Phagocytosis of Host Tissue by Invasive Malignant Cells

M. C. Van Peteghem

Department of Experimental Cancerology, Clinic for Radiotherapy and Nuclear Medicine, Academic Hospital, De Pintelaan 135, B-9000 Ghent, Belgium

ABSTRACT

Phagocytosis of fragments of host tissue by invasive malignant cells is demonstrated by confronting hypoblast cells, containing yolk droplets as intracellular markers, with malignant cells in three-dimensional culture. The yolk droplets appear during cultivation in the phagosomes of the malignant cells, while the hypoblast cells progressively disappear. The culture system consists of allowing adhesion of hypoblast fragments to spheroidal aggregates of malignant cells on a non-adhesive substrate, and transferring the confronting tissues to shaker culture. When hypoblast cells are confronted with malignant cells during spreading in two-dimensional monolayer culture, destruction of the hypoblast or yolk droplets inside the malignant cells are never observed. Time lapse cinematography of these monolayer confrontations nevertheless clearly demonstrates a distinct behaviour of hypoblast towards malignant as compared to non-malignant cells. The spreading hypoblast withdraws when coming into contact with a spreading explant of malignant cells, whereas it goes on spreading when confronted with non-malignant cells.

KEYWORDS

Invasion; phagocytosis; hypoblast; three-dimensional culture; two-dimensional culture; destruction; spheroidal aggregates; monolayer culture; in vitro cell interaction; malignant cells.

INTRODUCTION

Destruction and phagocytosis of host tissues by malignant cells are presumed to be involved in invasion. Since most malignant cells contain fairly large numbers of autophagosomes, recognition of heterophagocytosis in these cells is difficult. We have tried to solve this problem by confronting malignant cells with a host tissue that has an intracellular marker, which is absent from malignant cells and which can easily be recognized when inside another cell.

In the young chick blastoderm most cells hold a large number of yolk droplets, which are obvious both light microscopically and ultrastructurally and therefore may serve as a marker. Chick blastoderms have already been used for the study of malignant cells (Mareel, Vakaet and De Ridder, 1968; Sherbet and Lakshmi, 1978), and it appears that the hypoblast of such blastoderms is able to distinguish between

malignant and non-malignant cells (Mareel and co-workers, 1975). The hypoblast is easy to handle as a host tissue since it can be removed from the blastoderm and brought into culture separately (Sanders, Bellairs and Portch, 1978).
In order to study the capacity of malignant cells to phagocyte material from normal tissues we have confronted the chick hypoblast with malignant cells in both three-dimensional and two-dimensional culture. The confronting tissues are observed by stereomicroscopy, cinemicrophotography, light- and electronmicroscopy.

CONFRONTATION IN THREE-DIMENSIONAL CULTURE

MO_4 Cells and HeLa Cells

Spheroidal aggregates of MO_4 cells (malignant mouse fibroblastic cells) are prepared from trypsinized monolayers as described by Mareel, Kint and Meyvisch (1979). Aggregates of HeLa cells are made in the same way, but fragments of monolayers are used instead of cell suspensions. Hypoblast fragments are dissected from stage 4 (Vakaet, 1970) chick blastoderms. The confronting tissues are incubated consecutively on semi-solid medium and in shaker culture as described by Mareel and co-workers in this volume. Adhesion between both tissues is rather slow and necessitates 24 h incubation on the static semi-solid medium before transfer to the shaker culture.
Daily stereomicrography of individual cultures demonstrates a progressive increase in the volume of aggregates of MO_4 cells or HeLa cells and a decrease of the amount

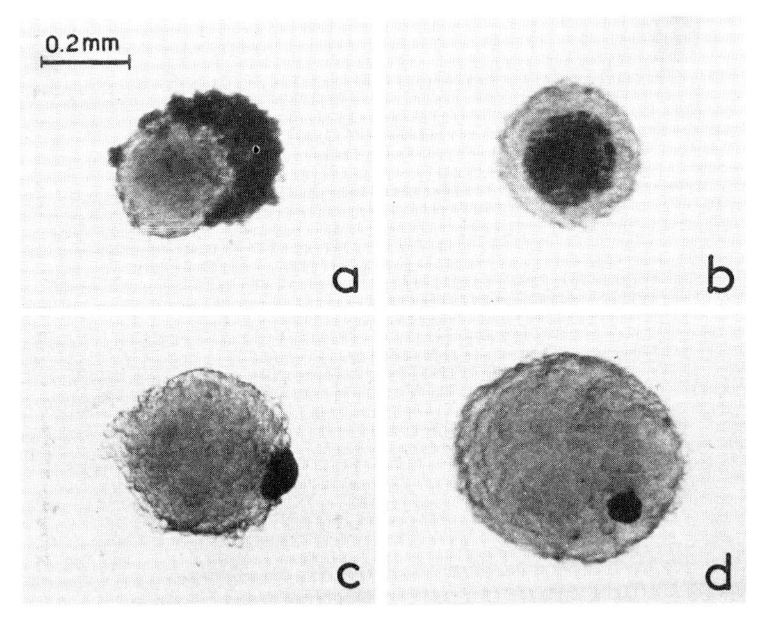

Fig. 1. Stereomicrographs of a confrontation between an aggregate of MO_4 cells and a fragment of chick hypoblast, after 1 day on semi-solid medium (a), after 1 (b), 2 (c) and 3 (d) additional days in shaker culture. The presence of yolk makes the hypoblast less translucent.

of hypoblast (Fig. 1). At the beginning of the shaker culture the hypoblast is flattened along the confronting aggregate. One (with HeLa) to 2 (with MO_4) days later, it has formed a nodule which is situated at one pole of the aggregate. After 5 (with MO_4) to 6 (with HeLa) days the hypoblast can no longer be recognized.

In order to see how far loss of material from the hypoblast into the medium is responsible for its disappearance, confrontations are kept on the semi-solid medium and followed by time lapse cinematography. Although some shedding of material is observed, it is obvious that the amount of hypoblast lost in this manner cannot account for its disappearance.

The histology of 2 µm thick sections from confrontations fixed after 1 to 5 days of incubation confirms the stereomicroscopic findings. The area of immediate contact between the aggregate and the hypoblast is limited to a few cells with HeLa, and more extensive with MO_4 aggregates. In this area MO_4 or HeLa cells are polarized; their fine cytoplasmic extensions point towards the hypoblast (Fig. 2). Necrotic cells and debris from the hypoblast are observed in the area of contact. Vacuoles containing yolk droplets are visible in MO_4 (Fig. 3) and HeLa cells. They are more numerous and appear earlier in MO_4 than in HeLa, where a lot of the yolk containing material is located between the cells. MO_4 or HeLa cells are not observed inside the bulk of the hypoblast.

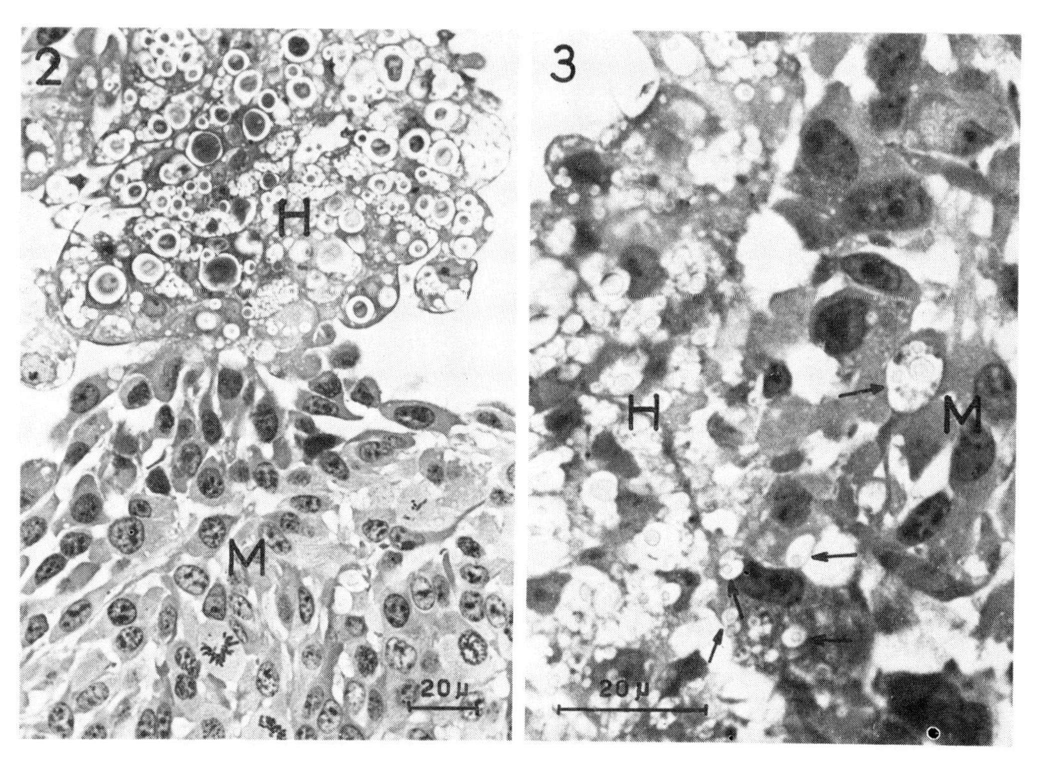

Fig. 2. Light micrograph of a 2 µm thick section from a confrontation of an aggregate of MO_4 cells (M) and a fragment of chick hypoblast (H), fixed after 1 day in shaker culture.

Fig. 3. Light micrograph of a section from a confrontation similar to that of Fig. 2. Arrows indicate vacuoles with yolk.

Transmission electron micrographs of confrontations of hypoblast with aggregates of HeLa or MO_4 cells (Figs. 4 and 5) show large phagosomes inside the malignant cells. These phagosomes contain either yolk only, or yolk plus unidentified debris,

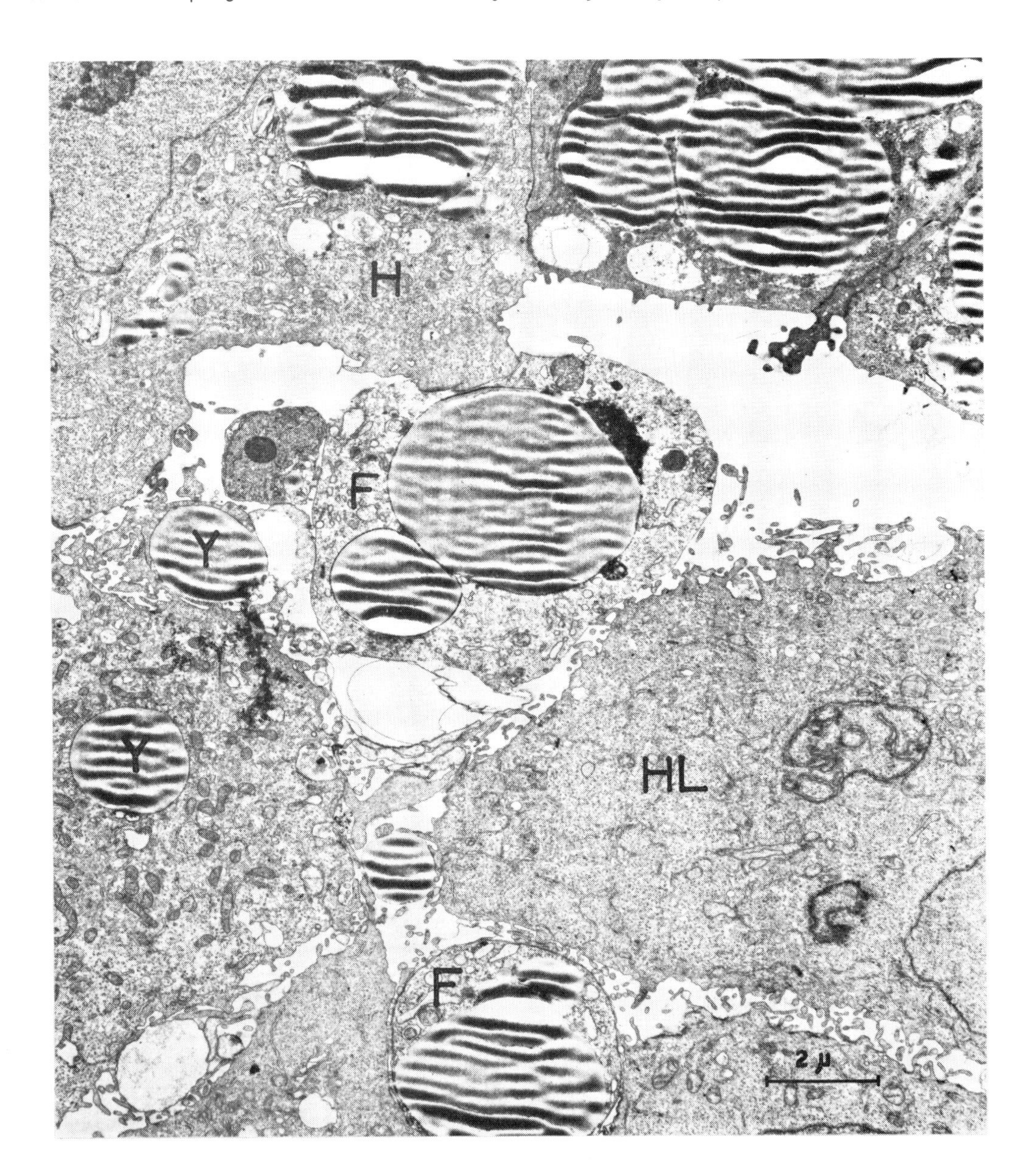

Fig. 4. Transmission electron micrograph of a confrontation between a fragment of hypoblast (H) and an aggregate of HeLa cells (HL) fixed after 3 days in shaker culture.
F : fragments from hypoblast·cells;
Y : yolk droplets, in phagosome and between cytoplasmic extensions at the periphery of HeLa cells.

Fig. 5. Transmission electron micrograph of a confrontation between a fragment of hypoblast (H) and an aggregate of MO_4 cells (M) fixed after 3 days in shaker culture. ⇉ : parallel apposition of plasmamembranes; ✱ : blebbing at periphery of hypoblast cell; → : cytoplasmic extension from MO_4 cell; Y : yolk inside MO_4 cell.

or debris only. Small yolk droplets are found in cells that are 4 to 5 cell diameters away from the area of immediate contact. Extracellularly located fragments of hypoblast cells are in contact with fine cytoplasmic extensions from the malignant cells. Hypoblast cells which are not in immediate contact with the confronting cells appear intact. Two types of contact between the hypoblast and the malignant cells are observed. The plasmamembranes of both confronting types of cells are parallelly apposed to each other. Or, fine extensions from the malignant cells penetrate between or through the hypoblast cells, similarly to the extensions from Novikoff hepatoma cells invading into the parenchyme of the liver in vivo (Babai and Tremblay, 1972). Such extensions are less frequent in HeLa than in MO_4 cells.

Hypoblast cells in immediate contact with malignant cells frequently show dilatation of the endoplasmic reticulum and irregular blebs at their periphery. It is tempting to speculate that these alterations of the hypoblast cell indicate a destructive activity of the malignant cells.

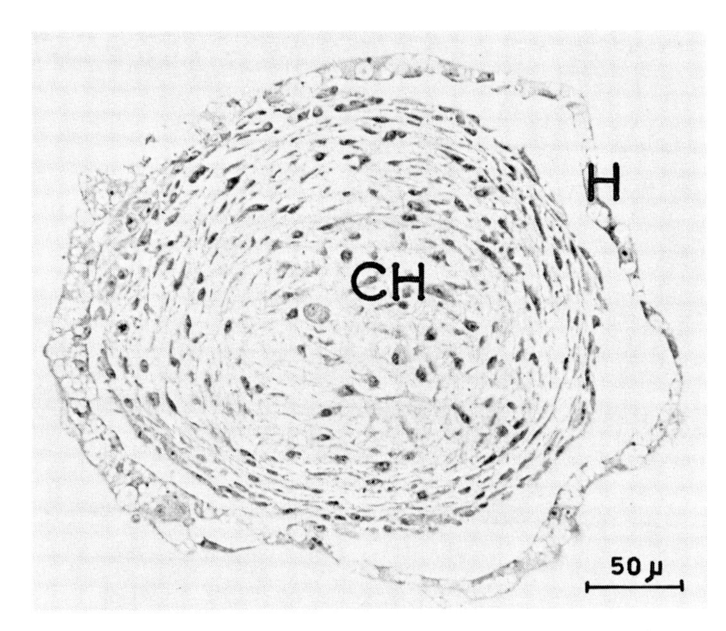

Fig. 6. Light micrograph of a section from a confrontation between a fragment of hypoblast (H) and an aggregate of embryonic chick heart cells (CH), fixed after 1 day in shaker culture. Compare with Fig. 2.

Embryonic Heart Cells

In order to see how far the fate of the hypoblast in the previous experiments is conditioned by its contact with the malignant cells, we have examined confrontations of fragments of hypoblast with aggregates of freshly prepared embryonic chick heart cells. In such confrontations the fragment of hypoblast rapidly surrounds the spheroidal aggregate of heart cells and forms a continuous epitheloid layer around it (Fig. 6). This chymearic structure can be kept in shaker culture for at least 3 days. Transmission electron micrographs from such structures reveal desmosomes between the hypoblast cells. The area of contact between the hypoblast and the aggregate of heart cells is small and shows parallel apposition of confronting plasmamembranes. Damage of hypoblast cells, or phagosomes containing yolk inside the heart cells, are not observed.

CONFRONTATION IN TWO-DIMENSIONAL CULTURE

When a fragment of hypoblast is explanted on glass or plastic, it rapidly spreads, presumably through the intense ruffling activity of the cells at its border, and covers an area of the substrate that has a diameter several times larger than that of the original explant. An aggregate of malignant cells (MO_4 or HeLa) or normal cells (embryonic chick heart or liver) put next to the fragment of hypoblast allows cinematographic analysis of the confrontation of cells migrating from both explants. Upon meeting with the advancing edge of an explant of malignant cells, the hypoblast withdraws (Fig. 7). This seems to be due to underlapping of the ruffling border of the malignant cells, which in the case of MO_4 cells eventually become located entirely between the substrate and the hypoblast. Total underlapping of HeLa cells was not observed. Despite the withdrawal, the ruffling activity of the hypoblast cells continues and hypoblast cells that are detached from the substrate are able to respread when coming into contact with a free area on the substrate. Contrary, spreading of the hypoblast continues upon meeting with cells migrating from aggregates of normal cells (Fig. 8). The latter have a tendency to revert their direction of movement. Underlapping of liver cells or of heart cells, or overlapping of hypoblast cells is occasionally observed at the common frontier. Transmission electron micrographs of confrontations between hypoblast and MO_4 or HeLa cells in two-dimensional culture are in agreement with the cinematographic observations. Cells of the spread hypoblast are connected by belt-desmosomes or show thin overlapping lamellae. The sites of contact with the artificial substrate are limited. MO_4 cells extend 0.1 µm thick filopodia underneath the hypoblast. When MO_4 cells are situated between the substrate and the hypoblast, parallel apposition of the plasmamembranes of both types of cells is observed. HeLa cells are never found between the substrate and the hypoblast; at their common frontier both cell types show broad contacts. At this site the hypoblast usually appears multilayered. Damage of hypoblast cells in contact with MO_4 cells or with HeLa cells, or phagosomes containing yolk inside the malignant cells, are not observed.

CONCLUSIONS

The presence of phagosomes which contain yolk inside MO_4 cells or HeLa cells shows that invasive cells are able to phagocyte material from the host tissue when confronted with it in three-dimensional culture. Degeneration of hypoblast cells in contact with the malignant cells suggests an immediate cytotoxic activity of the malignant cells. Further experiments, however, are needed to exclude other causes of degeneration of the hypoblast and to understand the mechanisms of the cytotoxic activity of malignant cells. Degeneration of hypoblast and phagocytosis of material from the hypoblast by the confronting malignant cells cannot be demonstrated in two-dimensional culture. Similar differences between both culture systems are described by Mareel, Kint and Meyvisch (1979) using embryonic chick heart as a host tissue. These observations indicate that confrontations in two-dimensional culture do not mimic invasion as well as three-dimensional confrontations. We think, however, that the two-dimensional cultures remain a valuable tool to study invasiveness of malignant cells for two reasons. First, comparison of the differential behaviour of host tissues in two- and three-dimensional culture might help us to understand how malignant cells destroy other tissues. Second, even in two-dimensional culture the hypoblast behaves differently when meeting malignant cells as compared to non-malignant cells.

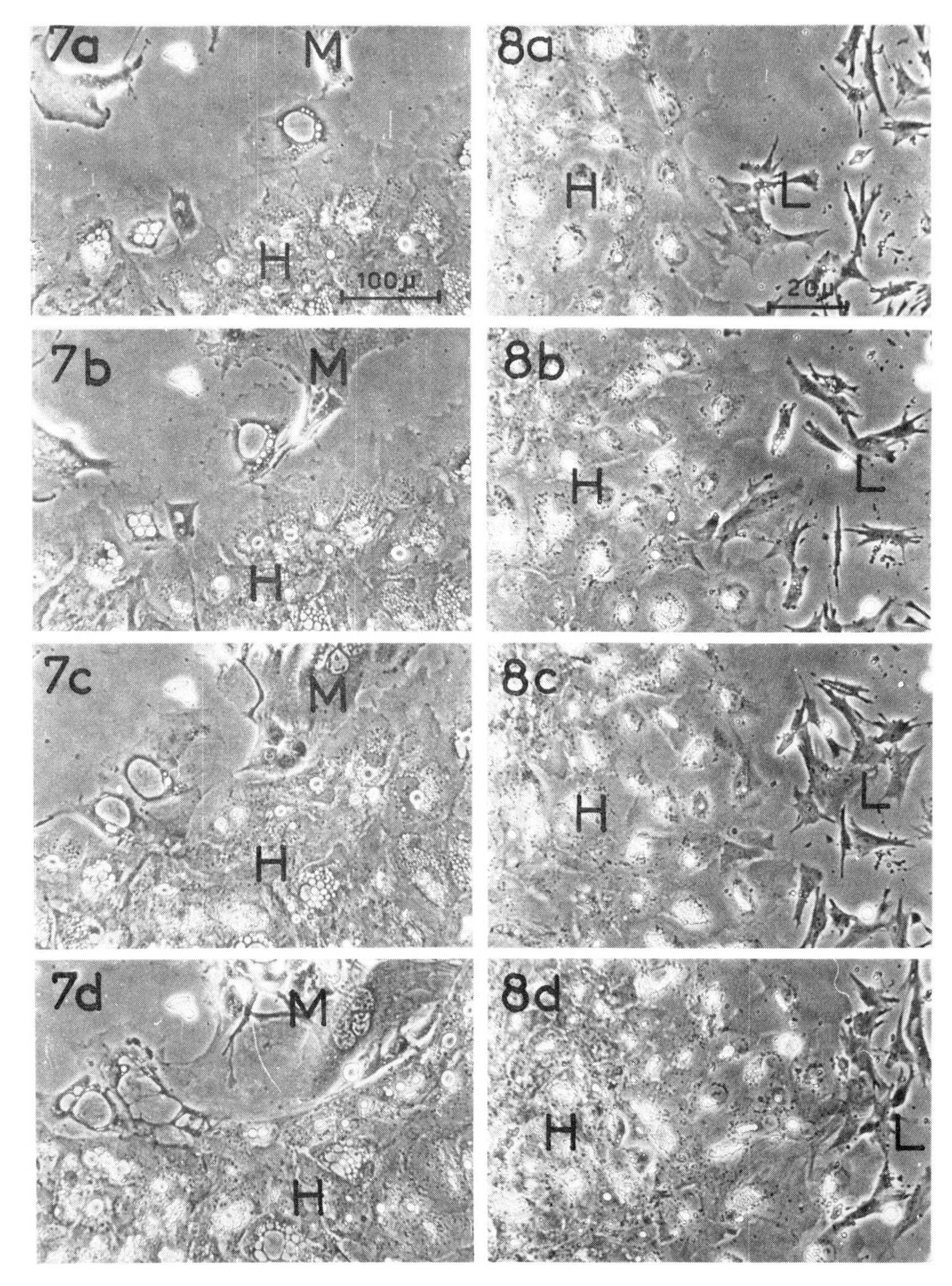

Fig. 7. Frames from a time lapse film of a fragment of hypoblast (H) confronting with MO_4 cells (M) migrating from an aggregate explanted on glass. a : 0 min; b : 2 h; c : 8 h; d : 18 h.

Fig. 8. Frames from a time lapse film of a fragment of hypoblast (H) confronting with liver cells (L) migrating from an aggregate explanted on glass. a : 0 min; b : 3 h; c : 5 h; d : 8 h.

ACKNOWLEDGMENT

This work is supported by a Grant from the Kankerfonds van de Algemene Spaar- en Lijfrentekas, Brussels, Belgium.

REFERENCES

Babai, F., and G. Tremblay (1972). Ultrastructural study of liver invasion by Novikoff hepatoma. Cancer Res., 32, 2765-2770.

Mareel, M., L. Vakaet, and L. De Ridder (1968). Grafting of HeLa cells in young chick blastoderms. Europ. J. Cancer, 4, 249-253.

Mareel, M., L. De Ridder, M. De Brabander, and L. Vakaet (1975). Characterization of spontaneous, chemical and viral transformants of a C3H/3T3-type mouse cell line through transplantation into young chick blastoderms. J. Natl. Cancer Inst., 54, 923-929.

Mareel, M., J. Kint, and C. Meyvisch (1979). Methods of study of the invasion of malignant C3H-mouse fibroblasts into embryonic chick heart in vitro. Virchows Arch. B Cell Path., 30, 95-111.

Sanders, E.J., R. Bellairs, and P.A. Portch (1978). In vivo and in vitro studies on the hypoblast and definitive endoblast of avian embryos. J. Embryol. exp. Morph., 46, 187-205.

Sherbet, G.V., and M.S. Lakshmi (1978). Malignancy and prognosis evaluated by an embryonic system. Grading of breast tumours. Europ. J. Cancer, 14, 415-420.

Vakaet, L. (1970). Cinephotomicrographic investigations of gastrulation in the chick blastoderm. Arch. Biol., 81, 387-426.

In vitro Studies of the Invasiveness of Cultured Malignant, "Spontaneously" Transformed and Normal Human Bladder Cells

P. Don, J. Kieler* and M. Vilien*
Ben Gurion Univ., School of Med., Beersheba, Israel
**The Fibiger Laboratory, Copenhagen O, Denmark*
Bladder Cells

ABSTRACT

The ability of 7 human cell lines to invade into cultured embryonic chick heart fragments was investigated. The 3 cell lines HU 456, HU 961 and T24 were derived from human bladder tumors of various degrees of malignancy. The 2 cell lines HCV29 and HU 609 were cultured from normal bladder and ureter epithelium, respectively. The sublines, HCV29T and HU 609T, derived from HCV29 and HU 609, showed "spontaneous" morphological transformation. The malignant and morphologically transformed cells invaded the heart fragments while the cells of normal origin apparently did not.

KEYWORDS

Invasion; cell lines; human bladder epithelia; three-dimensional culture.

INTRODUCTION

The invasive properties of malignant mammalian cells have been investigated in various in vitro systems. This has been studied with chick embryo mesonephros (Barski, and Wolff, 1965) or blastoderm (Mareel, and colleagues, 1975) serving as host tissue. Human embryos (Leighton, 1954), human decidua graviditatis (Schleich, Frick, and Mayer, 1976) and mouse mesonephros (Wolff, 1967) were some of the receptor tissues used to demonstrate the invasiveness of malignant human cells; for example tumors of the breast (Armstrong, and Rosenau, 1978) and intestine (Sigot-Luizard, 1974) and tissue culture lines such as HeLa (carcinoma), AFi (fibrosarcoma) and "spontaneously" transformed lymphoblastoid (Schleich, Frick, and Mayer, 1976).

Recently, Mareel, Kint, and Meyvisch (1979a) and De Ridder, Mareel, and Vakaet (1977) devised a simple, reproducible organotypical technique which involves the confrontation of spheroid aggregates of malignant mouse cells with fragments from embryonic chick heart. In collaboration with Dr. Mareel, we have applied this method to the study of invasion of "spontaneously" transformed ST/a mouse cells into chick heart tissue (Kieler, and colleagues , 1979). Presently, we are testing the invasive properties of human cells from various origins.

This assay of invasion is being investigated with the hope that it will be useful as a criterion of malignant alteration in our study of the influence of chemical carcinogens on human cells *in vitro*. This communication represents a preliminary report of the work we have done with 7 cell lines derived from human bladder and ureter cells.

MATERIAL AND METHODS

Cell Lines

The characterization of the cultured cells is summarized in Table 1.

TABLE 1 Cultured Cells

Cell line	Origin & Classification	Diagnosis & Clinical course	Chromosome no. Mode/Median	Growing tumors in nude mice
HU 456	TCC* Papillomatosum Non-infiltr. T_2, GRI	Died with invading tumor and metastasis of TCC origin	80/82	+
HU 961	TCC Papillomatosum Non-infiltr. T_1, GRII	Operated within a ½ year for recurrent bladder tumors,thereafter no recurrences	N.D.**	N.D.
T24	TCC Papillomatosum GRIII		91/91	+
HU 609	Ureter epithelium Normal microscopy	Kidney was removed due to hypernephroma, no metastasis and no recurrences	61/61	O
HCV29	Bladder epithelium Normal microscopy	Patient was irradiated for TCC of the bladder	46/45	O

*TCC - Transitional cell carcinoma
**N.D.- Not done.

The HCV29 was established in 1971 by Fogh (Bean and colleagues, 1974) at the Sloan Kettering Institute. The T24 was explanted by

Bubenik, and colleagues , (1973) at the Wenner-Gren Institute, Stockholm and propagated at the Institute of Experimental Biology and Genetics, Prague. These were obtained by the courtesy of Dr. Fogh. The HU 456, HU 961 and HU 609 were established in 1975-76 at the Fibiger Laboratory by Vilien. In the Wenner-Gren Institute and the Fibiger Laboratory at about passages 28-35, HU 609 and HCV29, "spontaneously" showed morphological transformation and were then classified as HU 609T and HCV29T. The 3 cell lines HU 456, HU 961 and T24 we classify as originally known malignant cells and the 2 cell lines HU 609 and HCV29 as being normal when first cultured in vitro.

The established cell lines were stored in liquid nitrogen. They have been propagated in vitro for more than 2 years through non-continuous weekly subcultivation in T75 vessels (NUNC, Roskilde) containing 10 ml Fib 41B which is a modification of MEM (Eagle, 1959) with a two-fold concentration of the essential amino acids, a four-fold concentration of the vitamins and of glutamine, 10% fetal calf serum (FCS) and penicillin and streptomycin.

Cocultures

The invasiveness of these 7 cell lines was studied by methods which were essentially the same as those originally described by Mareel (1979) except for minor modifications:

1) Cell suspensions were obtained by trypsinization or by scratching with a rubber policeman. The spheroid aggregates were prepared by incubation of a suspension of cells at $3\text{-}5 \times 10^5$/ml.

2) Cell aggregates, chick embryo heart fragments and cocultures were incubated in tightly stoppered flasks containing modified MEM with Earle's salts, 1/3 normal amount of $NaHCO_3$ (0,7g/l), plus 10% F.C.S. and not in continuous gas flow of 5% CO_2 plus 95% air. In order to maintain the proper pH, the media was either pregassed with CO_2 or mixed with Hepes buffer (20 mM).

3) Besides the standard hematoxylin and eosin (H. and E.) stain some paraffin sections were stained with phosphotungstic acid and hematoxylin (PTAH). The latter stain was useful in differentiating between chick heart and human cells.

4) Cell aggregate sizes ranged from 0.5-0.7 mm. and were as large as or bigger than heart fragments.

In reference to the incubation periods, day zero (D_o) refers to the day the heart and aggregate were cocultured in order to start the study of invasiveness; D_1 is 24 hours after cocultivation, D_2 is 48 hours etc.

RESULTS AND DISCUSSION

PTAH staining of aggregates showed darkly stained, large nuclei with almost non-stained cytoplasm. On the other hand, chick heart fragments had small darkly stained nuclei with a pinkish-colored cytoplasm.

Heart fragments from D_0 and D_{13} showed comparable histology and were surrounded by a layer of flattened cells Fig. 1, as described by Mareel, Kint, and Meyvisch (1979a). Spheroid aggregates were formed with all 7 human cell lines tested. These aggregates adhered well to the heart fragments Fig. 2.

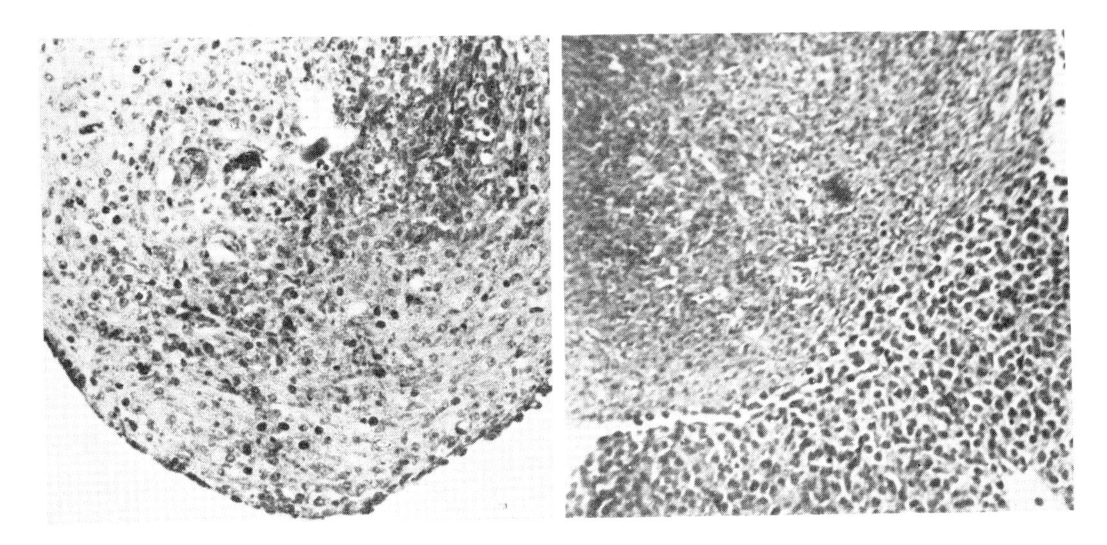

Fig. 1. Heart fragment at D_{13}. H. and E. x 870

Fig. 2. Association of heart fragment with T24 aggregate at D_o. H. and E. x 540

Replacement of the heart fragment by the rapidly invading HU 961 was essentially complete within 48 hours Fig. 3. A small necrotic area (N) is the only remnant of the heart visible and it persisted througout the incubation period. This is in agreement with the observation that invading cells circumvent necrotic areas (Latner, Longstaff, and Pradham, 1973). Figures 4 and 5, show the pattern of invasion of HU 456 and T24. Deadhesion of single cells from the aggregate bulk can be seen. Their penetration into the chick heart as solitary cells is indicative of malignancy (De Ridder, Mareel, and Vakaet, 1977). With time, the tumor cells progressively replaced the normal tissue. Outgrowth of tumor cells was observed when cocultures of heart fragments and tumor cells were explanted in tissue culture vessels.

Figures 6 and 7, show the invasion of the "spontaneously" morphologically transformed cells, HCV29T and HU 609T. These cell lines penetrated as a bulk into the host tissue Fig. 6, but deadhesion and penetration of single cells was also seen Fig. 7.

Aggregates of HCV29 and HU 609, cells of normal origin, adhered to heart fragments as well as malignant cells. Figure 8 shows that chick heart tissue encircled the HCV29 aggregate within 5 days of coculture. The region of contact between the 2 tissues is a well defined border. A comparable histological pattern was observed even after 13 days of incubation. Such a confrontation between non-malignant tissue fragments has been described by Mareel (1979b).

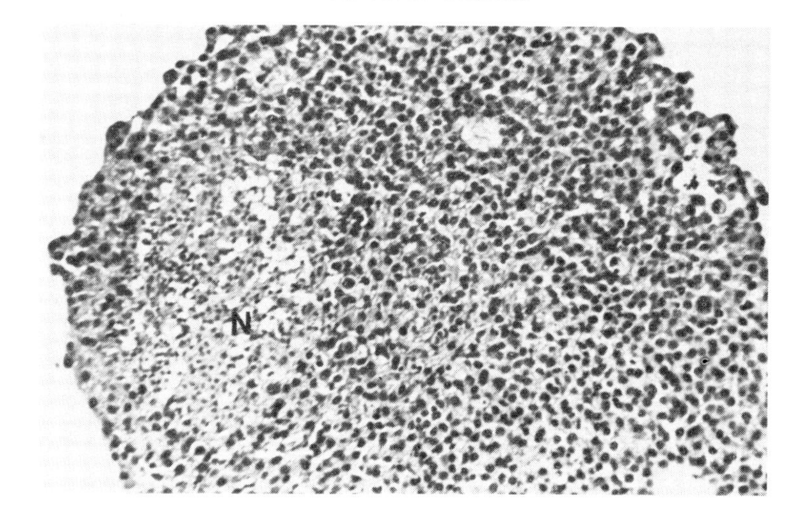

Fig. 3. Destruction of heart fragment at D_2 by rapidly invading HU 961 cells of TCC, grade II origin. N = necrotic remnant of heart tissue. H. and E. x 540

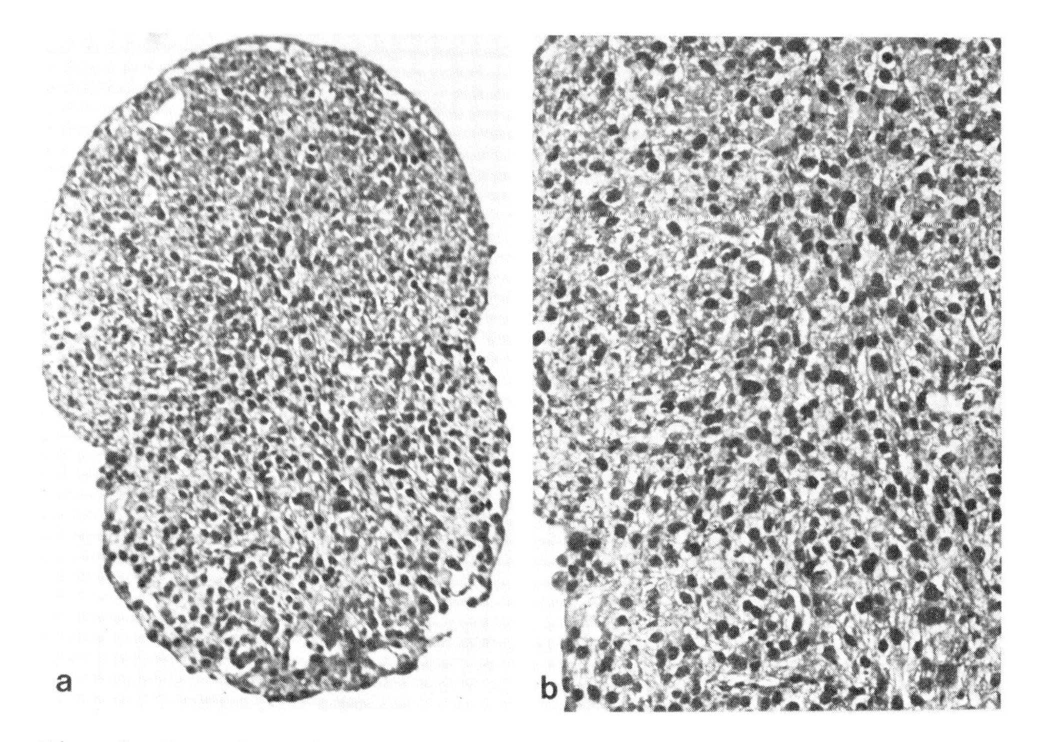

Fig. 4. Invasion of heart tissue (top) at D_3 by HU 456 cells (bottom) of TCC, grade I origin. H. and E. a: x 540. b: x 870

Fig. 5. Invasion of heart tissue (bottom, left) at D_3 by T24 cells (top, right) of TCC, grade III origin. PTAH x 870

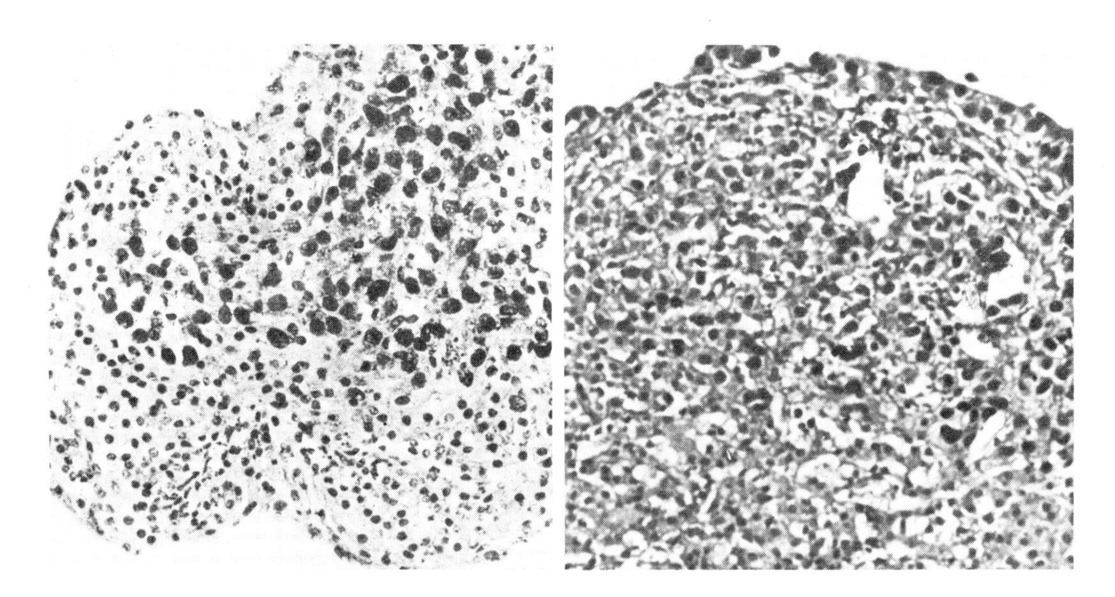

Fig. 6. Penetration of in vitro transformed HCV29T cells (top, right) of normal origin into heart tissue (bottom, left) at D_3. PTAH x 870

Fig. 7. Invasion of in vitro transformed HU 609T cells (top) of normal origin into heart tissue (bottom) at D_4. H. and E. x 870

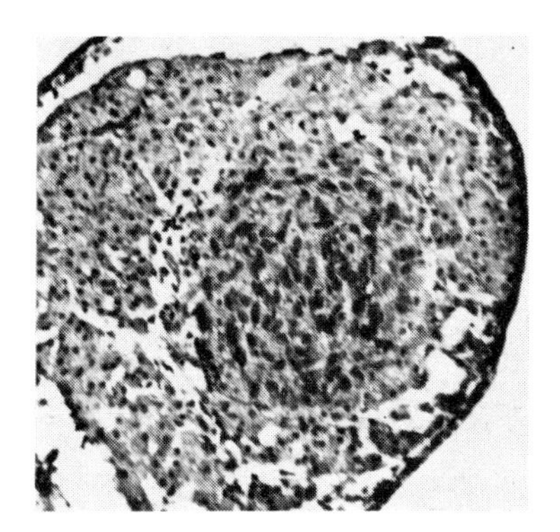

Fig. 8. Encirclement of HCV29 by heart tissue at D_5. H. and E. x 540

On D_0, the HU 609 aggregate adhered to the embryonic chick heart. With incubation, the aggregate appeared to fuse with the heart fragment at the original site of contact Fig. 9. A plane of interzone could be observed Fig. 10, between the 2 confronting tissues with perhaps some intermingling of cells from both tissues even after 10 days of coculture.

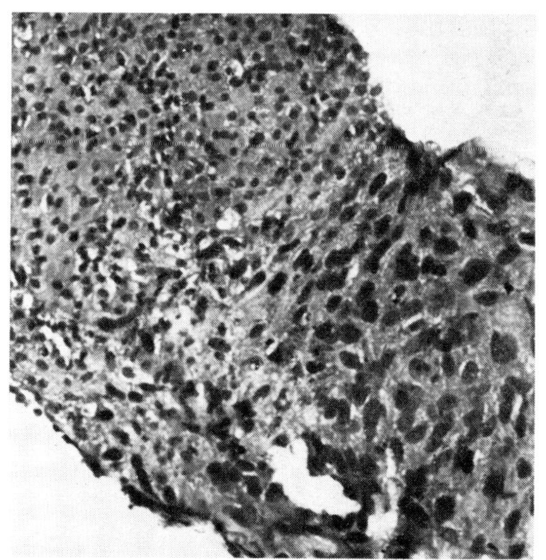

Fig. 9. Fusion of HU 609 cells of normal origin (right) with heart tissue (left) at D_4. H. and E. x 870

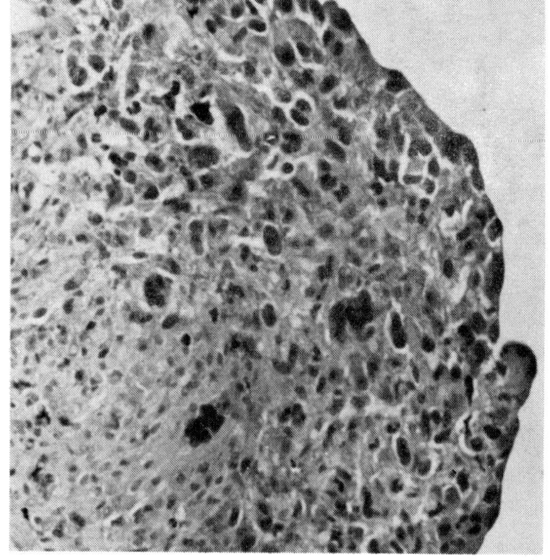

Fig. 10. Intermingling of HU 609 cells of normal origin (right) with heart cells (left) at D_{10}. H. and E. x 870

This has been a report of our pilot experiments and modifications are required. For example, we have found that Hepes buffer (20 mM) does not offer optimal coculture conditions. Malignant C3H mouse fibroblasts incubated in Hepes buffer invade into chick heart tissue. However, the pattern of invasion in Hepes buffer was different, not as complete or massive, when compared to cocultures maintained in a continuous gas flow of 5% CO_2 plus 95% air. Hepes buffer caused wide spread vacuolization of chick heart cells. This was observed in H. and E. stained paraffin sections and in outgrowths of chick heart explants grown in Hepes buffer. Electron microscopy of chick heart fragments cultured in Hepes buffer Fig. 11, showed that the cells contained large vacuoles, degenerating mitochondria, and a substantial decrease in glycogen and myofibril content.

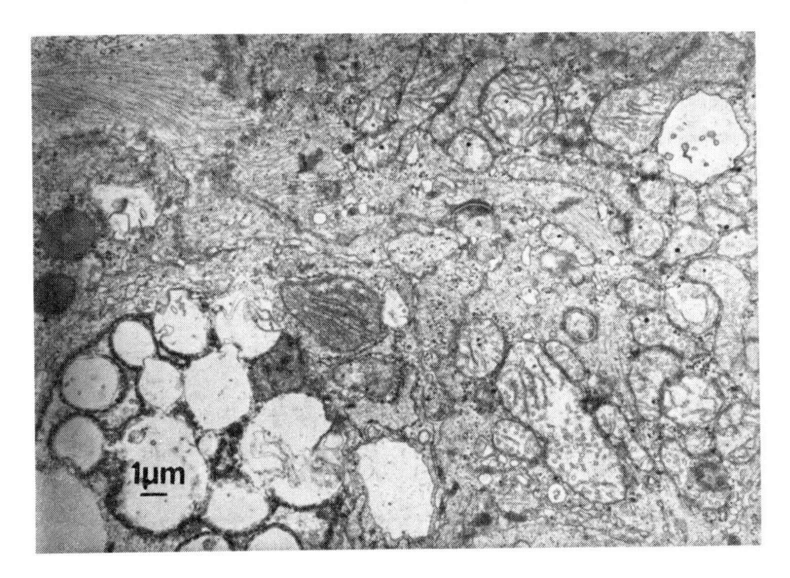

Fig. 11. Electron microscopy of heart fragments incubated in Hepes buffered MEM media, D_7.

In conclusion, we would like to stress once more that what we have presented are only preliminary results. In reality, until we have truly normal human bladder cell controls, we can't state how this assay of malignancy compares to other known criteria.

The question relative to us remains, if we use this technique to study the effect of carcinogens on human bladder cells _in vitro_, can we test the carcinogens on established aneuploid cell lines or must we use diploid early cultured cells? If the latter is the answer, new and reproducible techniques are needed to establish normal bladder epithelial cell lines.

ACKNOWLEDGEMENT

The authors would like to express their gratitude to Jacqueline Schmidt who assisted in the laboratory studies. The critical advice of Dr. M. Mareel is gratefully acknowledged. This work was carried out at the Fibiger Laboratory which is sponsored by the Danish Cancer Society.

REFERENCES

Armstrong, R.C., and W. Rosenau (1978). Cocultivation of human primary breast carcinomas and embryonic mesenchyme resulting in growth and maintenance of tumor cells. Cancer Res., 38, 894-900.
Barski, G., and E. Wolff (1965). Malignancy evaluation of in vitro transformation of mouse cell lines in chick mesonephros organ cultures. J. Natl. Cancer Inst., 34, 495-510.
Bean, M.A., H. Pees, J.E. Fogh, H. Grabstald, and H.F. Oettgen (1974) Cytotoxicity of lymphocytes from patients with cancer of the urinary bladder: Selection by a ^{3}H-proline microcytotoxicity test. Int. J. Cancer, 11, 186-197.
Bubenik, J., M. Baresova, V. Viklicky, J. Jakoubkova, H. Sainerova, and J. Donner (1973). Established cell line of urinary bladder carcinoma (T24) containing tumour-specific antigen. Int. J. Cancer, 11, 765-773.
De Ridder, L., M. Mareel, and L. Vakaet (1977). Invasion of malignant cells into cultured embryonic substrates. Arch. Geschwulstforsch., 47, 7-27.
Eagle, H. (1959). Amino acid metabolism in mammalian cell cultures. Science, 130, 432-437.
Kieler, J., P. Briand, M.C. Van Peteghem, and M. Mareel (1979). Comparative studies of two types of "spontaneous" malignant alteration of ST/a mouse lung fibroblasts propagated in vitro. In Vitro. Accepted for publication.
Latner, A.L., E. Longstaff, and K. Pradham (1973). Inhibition of malignant cell invasion in vitro by a proteinase inhibitor. Brit. J. Cancer, 27, 460-464.
Leighton, J. (1954). Studies on human cancer sponge matrix tissue culture. I. The growth patterns of a malignant melanoma, adenocarcinoma of the parotid gland, papillary adenocarcinoma of the pancreas and epidermoid carcinoma of the uterine cervix (Gey's HeLa strain). Tex. Rep. Biol. Med., 12, 847-864.
Mareel, M., L. De Ridder, M. De Brabander, and L. Vakaet (1975). Characterization of "spontaneous" chemical and viral transformants of a C3H/3T3 type mouse cell line by transplantation into young chick blastoderms. J. Natl. Cancer Inst., 54, 923-929.
Mareel, M., J. Kint, and C. Meyvisch (1979a). Methods of study of the invasion of malignant C3H-mouse fibroblasts into embryonic chick heart in vitro. Virchows Arch. B. Cell Path., 30, 95-111.
Mareel, M.M.K. (1979b). Is invasiveness in vitro characteristic of malignant cells? Cell Biology. Accepted for publication.
Schleich, A.B., M. Frick, and A. Mayer (1976). Patterns of invasive growth in vitro. Human decidua graviditatis confronted with established human cell lines and primary human explants. J. Natl. Cancer Inst., 56, 221-237.

Sigot-Luizard, M.F. (1974). The association of tumor and embryonic cells in vitro. In G.V. Sherbert (Ed.), Neoplasia and Cell Differentiation, Karger, Basel. pp. 350-379.

Wolff, E. (1967). Le mecanisme de l'invasion du cancer en culture organotypique. In P. Denoix (Ed.), Mechanisms of Invasion in Cancer, Springer-Verlag, New York. pp. 204-211.

Invasiveness of Neutrophil Leukocytes

Peter B. Armstrong

Department of Zoology, University of California, Davis, California 95616, U.S.A.

ABSTRACT

Intercellular invasion can be defined as the penetration of cells into the interior of a dissimilar tissue. A variety of cells, including malignant cells, motile blood cells and a number of embryonic cell types are invasive. For many cells, invasion is dependent on active cellular locomotion. A number of mechanisms potentially important for invasion were examined for neutrophil leukocytes using *in vitro* and *in vivo* systems. Evidence is presented that invasion does not depend on tissue growth, destruction of the host tissue, defective cellular adhesive recognition, release of serine proteases or ability to respond to chemotactic stimuli. Factors of potential importance displayed by neutrophil leukocytes include low adhesiveness, rapid locomotion, high deformability and an absence of contact paralysis of pseudopodial activity during interaction with cells of the host tissue.

KEYWORDS

Cellular motility; intercellular invasion; contact inhibition of locomotion; tissue recognition; cellular adhesion; neutrophil leukocyte; tissue stability; chemotaxis; morphogenetic movement; protease inhibitors.

INTRODUCTION

Intercellular invasion can be defined as the penetration of cells of one type into the interior of other tissues (Abercrombie, 1970). Invasion contrasts markedly with the behavior of the cells of most of the coherent tissues of the body which, instead of spreading into adjacent tissues, remain confined within the tissue boundaries for the lifetime of the organism. The mechanisms responsible for stabilizing tissue architecture (Table 1) are of interest since most tissue cells are potentially motile, as shown by their behavior during wound healing (Lash, 1955; Radice, 1979) and when explanted into tissue culture (Abercrombie, 1973; Goldman, Pollard and Rosenbaum, 1976).

INVASIVE CELL TYPES

A variety of cell types are invasive. The invasiveness of malignant cells represents a major mechanism of tumor dissemination (Easty, 1966; Fidler, 1975; Sträuli

TABLE 1 Mechanisms Potentially Responsible for Stabilizing Tissue Architecture

Mechanism	References
Contact inhibition of locomotion: Pseudopodial activity is suppressed at sites of cell-cell contact	Abercrombie (1970)
Reversible loss of components necessary for motility in quiescent tissue cells	Gabbiani (1979a); Lefford (1972)
Tissue cells adhere with junction too strong to be broken by the forces generated by cell locomotion	Weinstein, Merk and Alroy (1976)
Impenetrable intercellular matrices immobilize tissue cells	
Mechanical barriers to movement such as the basement membrane delineate tissue interfaces	Kefalides (1978); Vracko (1974)
Tissue recognition prevents intermingling of cells of contiguous tissues	Fig. 1; Steinberg (1970)
Non-adhesive plaques at the outer faces of epithelia prevent cells from spreading over the surfaces of epithelia	Buck (1973): DeRidder, Mareel and Vakaet (1975); Di Pasquale and Bell (1974); Roberson and Armstrong (1979)

and Weiss, 1977; Willis, 1960). Most of the motile blood cells of vertebrates (eg. neutrophils (Marchesi, 1970; Grant, 1973), eosinophils (Archer, 1970; Spiers and Osada, 1962), macrophages (Speirs, 1970), and lymphocytes (Schoefl, 1972)) are able to move across the vascular endothelium to enter and invade the subjacent connective tissue. Invasion also occurs in a number of situations during embryonic morphogenesis. The superficial layer of the early mammalian embryo, the trophoblast, in many species is highly invasive to the wall of the uterus during implantation (Blandau, 1961; Boyd and Hamilton, 1952; Schlafke and Enders, 1975; Sherman and Wudl, 1976). Neural crest (Johnston and Listgarten, 1972; Norden, 1973; Twitty, 1949), primordial germ cells (Blandau, White and Rumery, 1963; Witschi, 1948), and nerve axon growth cones (Speidel, 1964) are invasive during embryogenesis. The presumptive hepatocytes invade the mesenchyme of the ventral mesentery during liver morphogenesis (Rifkind, Chin and Epler, 1969), and cardiac endothelium invades the wall of the ventricle during heart development (Manasek, 1970; Patten, Kramer and Barry, 1948). In the adult, certain neural crest derivatives (viz., dermal melanocytes (Billingham and Silvers, 1970) and capillary endothelial cells (Clark, 1912; Folkman and Cotran, 1976; Schoefl, 1964)) retain invasive activities.

Thus, it can be seen that although invasive behavior is limited to particular cell types, it is still a widespread phenomenon. It is found pathologically in malignant invasion and invasion of parasitic amoebae (De Lamater and others, 1954; Eaton, Meerovitch and Costerton, 1970; Neal, 1960) and normally in the infiltrative activities of white blood cells and a variety of embryonic cell types.

ROLE OF ACTIVE CELLULAR MOTILITY IN INVASION

Two basic categories of invasive processes can be envisioned: invasion dependent on active, pseudopod-directed cellular locomotion and invasion dependent on tissue growth or shape change and not involving cellular locomotion. This report concerns itself with motility-dependent invasion. The involvement of active locomotion has been demonstrated by direct observation of motility of cells invading suitably transparent host tissues (Table 2). The involvement of active motility in invasion is suggested by fine-structural observations which show protrusions of invading cells that resemble pseudopods and that penetrate between and indent the surfaces of cells of the host tissue (Ashworth, Stembridge and Luibel, 1961; Babai, 1976; Babai and Tremblay, 1972; Butterworth, 1970; Carr, McGinty and Norris, 1976; Chew, Josephson and Wallace, 1976; Enders and Schlafke, 1972; Fisher and Fisher, 1961; Locker, Goldblatt and Leighton, 1970; Luibel, Sanders and Ashworth, 1960; Marchesi, 1970; Mao, Nakao and Angrist, 1966; Nilsson, 1960; Ozzello and Sanpitak, 1970; Riley and Seal, 1968; Schoefl, 1964, 1972; Seal, Riley and Inman, 1969; Sugar, 1963; Tarin, 1967; Tickle, Crawley and Goodman, 1978b; Woods and Smith, 1969). Active

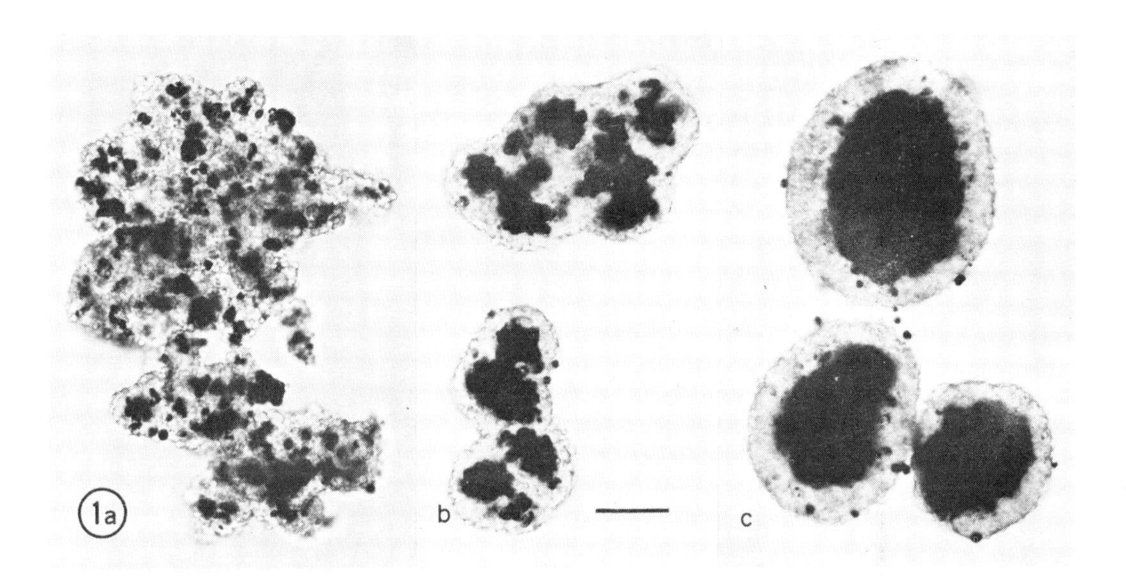

Fig. 1. The dynamic aspects of stabilization of cellular positioning in tissues is illustrated by the phenomenon of cell sorting, shown here in living aggregates comprised of a mixed population of chick embryo neural-retinal and pigmented-retinal epithelial cells. The former are clear; the latter are black. The mixed aggregates are prepared by allowing a mixed population of dissociated cells to reaggregate in stirred suspension culture. Five hours after reaggregation was begun, large aggregates have formed which contain both cell types, dispersed in a disordered and highly intermingled arrangement (1a). When these initially disordered aggregates are maintained in culture, the cells sort out into homogeneous tissue regions. Figure 1b was at 19 hours and shows an intermediate stage in the process. Nearly perfect sorting is evident at 2 days (1c), with the pigmented retina present as a single interior mass surrounded by a layer of neural retina. The cell sorting phenomenon demonstrates that, even when dissimilar tissue cells are comingled artificially, they have the capacity to move about to demingle into homogeneous tissues. In contrast, intercellular invasion results in the active intermingling of dissimilar cells. Bar = 100 µm.

locomotion is suggested also by those situations in which invasive cells have in histological preparations been observed as individual cells situated deep in the host tissue, separate from the parent mass of the invasive tissue (Armstrong and Lackie, 1975; Tickel, Crawley and Goodman, 1978a). In sum, there is strong evidence that invasion of a variety of cell types involves active, pseudopod-directed cellular locomotion. As has been pointed out previously, however, the evidence relating motility to invasion for solid tumors is less complete than might be wished (Strauli and Weiss, 1977; Weiss, 1977).

TABLE 2 Direct Microscopic Observation of Active Motility of Invading Cells in Life

Invading cell	Host tissue	Reference
Nerve axon, Schwann cell	Tadpole tailfin	Billings-Gagliardi (1977); Speidel (1964)
White blood cells	Tadpole tailfin	Clark, Clark and Rex (1936)
Neutrophil leukocyte	Chick embryo chorioallantoic membrane	Armstrong, unpublished
Leukocytes	Rabbit ear chamber	Allison, Smith and Wood (1955); Cliff (1966); Graham and others (1965); Grant, Palmer and Sanders (1962)
Yolk sac amebocyte	Teleost embryo	Armstrong (1979)
Capillary endothelium	Tadpole tailfin	Clark (1912)
Carcinoma cell	Rabbit ear chamber	Wood (1958); Wood, Baker and Marzocchi (1957, 1968)
Leukemia cell	Rat mesentery	Haemmerli and Strauli (1978)

DESTRUCTIVE VERSUS BENIGN INVASION

Invasion can occur in situations in which the host tissue suffers substantial histolysis or in situations in which host tissue destruction is minimal. The best studied example of host tissue destruction attendant with invasion occurs during invasion by solid tumors. Tissue destruction results in replacement of host tissue by tumor (Katsuda, Takaoka and Nagai, 1968; Leighton, 1967, 1968; Willis, 1960; Wolff, 1963). Invasion by pathogenic amoebae is generally accompanied by host tissue destruction (Culbertson, 1971; Cursons and Brown, 1978; Eaton, Meerovitch and Costerton, 1970). Destruction of uterine tissues accompanies invasion of the trophoblast in mammals showing deep implantation (Schlafke and Enders, 1975) and tissue destruction is very marked during trophoblastic invasion at ectopic sites (Kirby, 1965). During embryonic development, cell death is an important process in morphogenesis (Saunders, 1966) and sometimes accompanies invasion. For example, in avian embryos, the migration of the primordia of the sternum is accompanied by extensive cell death of mesenchymal cells in advance of the migrating rudiment (Fell, 1939). Since the primordia are massive chondrified elements which are dragged along by the activities of fringes of undifferentiated motile cells, cell death may be required in this case to provide cell-free space into which the invading tissue can move.

In contrast to the examples of destructive invasion cited above, motile blood cells and leukemic cells appear able to penetrate the host tissue with minimal initial tissue destruction (Armstrong, 1977; Haemmerli and Strauli, 1978). Invasion by solid tumors is not invariably destructive (Wilson and Potts, 1970). Similarly, the invasion by neural crest and mesenchymal cells occurring during embryonic development appears not to involve destruction of host tissues (cf., Armstrong and Armstrong, 1978).

MECHANISMS OF INVASION

Since both noninvasive and invasive cells can engage in active pseudopod-directed locomotion, the capacity for motility alone cannot account for the differences in behavior between the two classes of cells. A variety of characteristics of the invasive cells can be envisioned as being of potential importance to the display of invasive behavior (Table 3). As can be seen by comparing Table 3 with Table 1,

TABLE 3 Mechanisms Potentially Responsible for Intercellular Invasion

Mechanism	References
Defective expression of contact inhibition of motility	Abercrombie (1970); Armstrong and Lackie (1975)
Contractile machinery is intact in invasive cells under conditions where it is incomplete in noninvasive cells	Gabbiani (1979b)
Adhesion of invasive cells is low; cell junctions are sparse in invasive cells	Coman (1944); Weinstein, Merk and Alroy (1976)
Invasive cells release lytic enzymes which disrupt the intercellular matrix and basement membrane	Reich (1978); Stickland, Reich and Sherman (1976)
Invasive cells display defective cellular recognition abilities	Coman (1961)
Chemostatic responsiveness is necessary for invasion	Dubois and Croisille (1970); Folkman and Cotran (1976); Romualdez and Ward (1975); Twitty and Niu (1954); Zigmond (1978)
Invasive cells destroy and phagocytose host tissue	See preceding section
Growth pressure forces invasive cells passively into the host tissue	Eaves (1973); Hamperl (1967); Young (1959)

many of the mechanisms of potential importance to invasion represent a breakdown of mechanisms of potential importance for tissue stabilization.

THE NEUTROPHIL LEUKOCYTE

Since the principal cell type whose invasive behavior is discussed in the succeeding sections is the mammalian neutrophil leukocyte, an introduction to this cell is in order. The neutrophil is the principal cell found in most types of the acute inflammatory response for the first 24-48 hours (Hirsch, 1973). During the initial stages of inflammation, circulating neutrophils adhere to (Atherton and Born, 1972) and then migrate across (Marchesi, 1970; Grant, 1973) the vascular endothelium of venules in the inflammed region. Once in the connective tissue subjacent to the venules, the cell functions as a phagocyte in the removal and destruction of invading bacteria and tissue debris. Neutrophils are produced and mature in the bone marrow and are released into the circulation as short lived cells incapable of cell division.

MECHANISMS OF INVASION BY NEUTROPHIL LEUKOCYTES

A number of descrete factors of potential importance to intercellular invasion can be identified (Table 3). An important goal of experimental studies of invasion is to provide an evaluation of the relative significance of each of these factors for the invasive behavior of each of the variety of invasive cell types. This section reports studies from my laboratory on the mechanisms underlying the invasive behavior of neutrophil leukocytes.

Ideally, studies of invasion of a particular cell type should include experimental systems ranging from tissue- and organ-culture models (useful for the ease of experimental manipulation) to systems of increasing complexity and including studies on the intact organism. The systems used for the study of neutrophil invasion include monolayer tissue culture alone or in combination with tissue cells, culture with solid aggregates of a potential host tissue in organ culture (Fig. 2a), explantation to the chick embryo chorioallantoic membrane (Fig. 2b) and observation of cells in the transparent tailfin tissue of tadpole larvae of the frog *Xenopus laevis*. Similar systems have been used by others to investigate malignant invasion (de Ridder, Mareel and Vakaet, 1977; Easty and Easty, 1963; Leighton, 1967; Schleich, Frick and Mayer, 1976; Wolff and Wolff, 1967). The studies have been designed to evaluate the importance of several of the mechanisms that potentially contribute to invasive behavior.

Cell motility versue growth pressure. Since the mature neutrophil is not capable of division, invasion by tissue growth can be ruled out. The cells are highly motile and highly deformable (Armstrong and Lackie, 1975; Wilkinson, 1978). Both factors almost certainly contribute to the ease with which neutrophils invade coherent tissues. The importance of cellular deformability is suggested by studies of Lichtman (1970) on leukocyte maturation. Maturation occurs in compartments in the bone marrow which are separated from the vasculature by diaphragms perforated by narrow fenestrae. The immature leukocytes are unable to crawl through the fenestrae, due apparently to their low deformability. Deformability increases markedly during maturation, resulting in cells which are eventually able to squeeze through the narrow gaps in the barrier to enter the circulation.

Cell adhesion. Neutrophils are definitely weakly adhesive, not only to other neutrophils, but also to other cell types and to inert surfaces. Although artificially disaggregated tissue cells are able to reestablish strong adhesive bonds with each other when placed under suitable culture conditions (Armstrong, 1966; Steinberg, Armstrong and Granger, 1973), neutrophils form loose clusters that are easily redispersed under conditions of mild shear. In monolayer culture, neutrophils crawling about on undersurface of a coverslip bearing a sparse layer of fibroblasts were frequently observed to lose their grip on the substrate (either glass or fibroblast) and to fall off (Armstrong and Lackie, 1975). Cells of coherent tissues of either epithelial or mesenchymal origin adhere tenaciously to the substrate under comparable culture conditions. Removal of cultured tissue cells requires strong shearing forces and usually results in fragments of cells being left behind still firmly attached to the substrate (Weiss and Coombs, 1963).

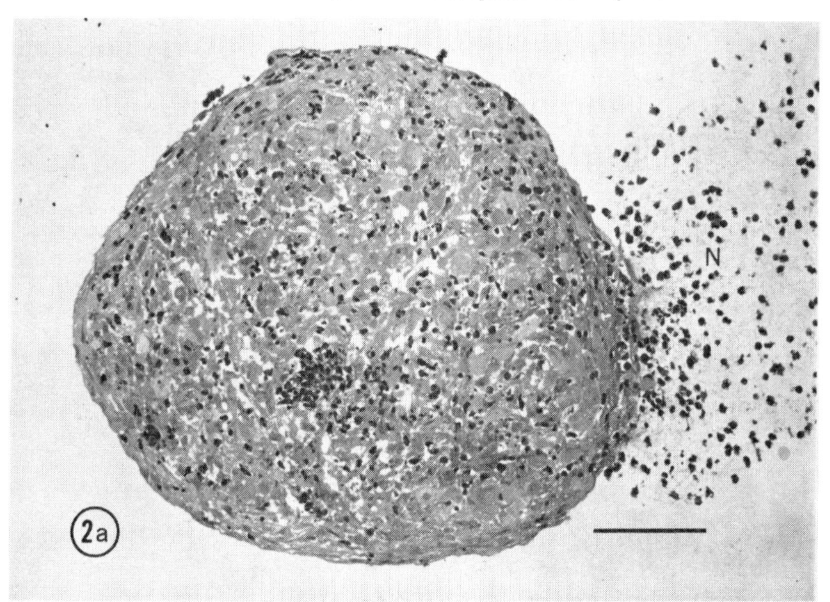

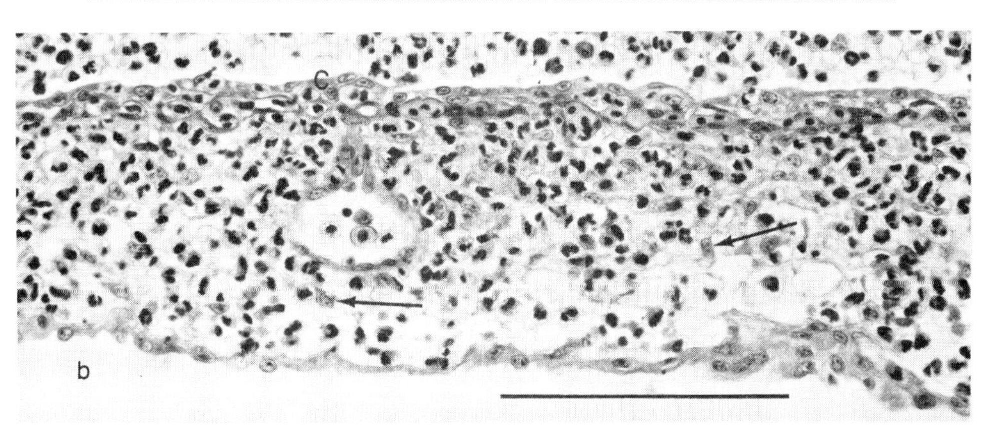

Fig. 2. A variety of model systems are useful in the experimental study of invasion. In Fig. 2a, neutrophils have invaded an aggregate of chick embryo heart fibroblasts in organ culture. The loose aggregate of neutrophils which was the source of invading cells is indicated by N. The dark-staining bodies in the interior of the fibroblast aggregate are the nuclei of neutrophils that have participated in invasion. Figure 2b shows the invasion of neutrophils placed initially on top of the chorioallantoic membrane of the 9-day chick embryo. Neutrophils have invaded across the chorionic epithelium (C) and into the subjacent mesenchyme, where they are present in very large numbers. Neutrophils have dark heterochromatic nuclei and are easily distinguished from cells of the host tissue (arrows indicate host tissue nuclei). In both cases,

invasion is rapid and massive (the tissues have been together for 2 hrs in 2a and for 1/2 hr in 2b), yet there is no sign of destruction of the host tissue. In 2b the chorionic epithelium is morphologically intact after having been traversed by large numbers of neutrophils. Bar = 100 μm.

In fine-structural studies, I have not found specialized junctions of the desmosone, gap or tight junction types established between neutrophils or between neutrophils and cells of a host tissue being invaded by neutrophils. In monolayer culture, neutrophils do not form specialized focal contacts of the type formed by fibroblasts (Armstrong and Lackie, 1975). Presumably the relatively low adhesiveness of neutrophils facilitates rapid locomotion and invasion. Since the cells do locomote readily, adhesion is strong enough to allow cells to gain purchase on the substrate but apparently is not so strong that the cells experience difficulty breaking adhesions during continued locomotion.

Lytic enzymes. It can be suggested that lytic enzymes might facilitate invasion either by removing elements of the extracellular matrix and basement membrane that impede invasion or by stimulating cells that in the absence of the enzymes would remain sedentary to adopt an invasive mode of behavior (Ossowiski and others, 1973). Neutrophils contain a variety of lytic enzymes, the best characterized of which are the proteases (Baggiolini, 1972; Janoff, 1972). Leukocytes release proteases into the external milieu, especially during periods of phagocytosis. (Leukocytes drool when they eat.) Of these, presumably proteases active at neutral pH, such as collagenase, elastase and plaminogen activator would be of greatest potential significance.

One way to test for the importance of proteases in invasion would be to study the invasive potential of leukocytes which lack the usual levels of proteases due to gene mutations present in the donor animal. The beige mouse and Chédiak-Higashi syndrome human represent possible sources of protease-deficient leukocytes (Vassalli and others, 1978). In my laboratory, we have investigated the effects of a variety of nontoxic protease inhibitors on invasion. Inhibitors included soybean trypsin inhibitor (10 mg/ml), ovomucoid inhibitor (10 mg/ml), PMSF (0.1 mM), TLCK (10 mg/ml) and TPCK (10 mg/ml). These levels of inhibitor did not suppress invasion. Unfortunately, none of the inhibitors used were effective against collagenase. Ar present, we are studying the effect of α_2-macroglobulin which is an inhibitor present in the serum that can suppress the proteolytic activities against macromoleculor substrates for a broad spectrum of endopeptidases.

Cellular recognition. The adhesive recognition abilities of tissue cells may contribute significantly to tissue stability (Fig. 1). Based on studies of transformed cells, it has been suggested that defective tissue recognition properties might be necessary for invasion to occur (Coman, 1961; Cassiman and Bernfield, 1976b; Dorsey and Roth, 1973; Nicolson and Winkelhake, 1975). Specific adhesive recognition can be studied by determining the relative rates of homotypic versus heterotypic adhesion in cultures containing two different cell types (Cassiman and Bernfield, 1976a; McGuire and Burdick, 1976; Moyer and Steinberg, 1976; Roth, McGuire and Roseman, 1971; Takeichi and others, 1979; Walther, Ohman and Roseman, 1973). Strong self-recognition is demonstrated by preferential establishment of homotypic contacts; weak self-recognition by the preferential establishment of heterotypic contacts. The discriminatory ability of neutrophils was assessed by allowing aggregation of dispersed cells to occur in stirred suspension culture. Cell suspensions contained 50% neutrophils and 50% fibroblasts (a tissue type that can be invaded by neutrophils - Fig. 2a). The aggregates that formed were scored microscopically for the number and identity of the cells present. The aggregates were strongly cell-type specific. Most aggregates contained only fibroblasts or only leukocytes and few aggregates contained both cell types (Lackie and Armstrong, 1975), indicating that adhesive recognition is intact in this system.

Chemotaxis. Neutrophil leukocytes possess a well developed ability to migrate in a directed fashion in response to chemotactic stimuli (Gallin and Quie, 1978; Wilkinson, 1974; Zigmond, 1978). The involvement of chemotaxis in the invasive behavior of neutrophils has been studied by determining the effect on invasion of agents known to suppress chemotactic migration of neutrophils. The drugs tested were the anti-inflammatory agents prednisolone (100 μg/ml), hydrocortisone succinate (100 μg/ml), ibuprotene (100 μg/ml), and naproxene (100 μg/ml) and the anti-microtubule agents colchicine (0.1 and 0.5 mM) and vinblastine sulfate (5μM). The numbers of invading neutrophils were determined by cell counts on sectioned preparations. Even at the high concentrations employed, none of the agents significantly reduced invasion. Thus, invasion *per se* appears not to require chemotactic responsiveness of the invading neutrophils. Chemotaxis may, however, play a directive role once invasion has begun. The insensitivity to anti-microtubule agents is interesting since these agents can suppress the invasion of tumor cells (Mareel and De Brabander, 1978) and of cardiac fibroblasts into myocyte aggregates (Armstrong and Armstrong, 1979)

Host tissue destruction. Based both on histological studies (at the light and electron microscope levels) and on observation of neutrophils migrating about in CAMs maintained in organ culture, I have been able to detect no sign of host tissue destruction during the first several hours of invasion by neutrophils (see Fig. 2b). The leukocytes appear to migrate between cells of the chorionic epithelium to gain access to the subjacent mesenchyme without causing observable damage to the host tissue.

Contact inhibition of motility. Since tissue cells are capable of active locomotion yet do not wander beyond the boundaries of the parent tissues, it can be suggested that the inherent capabilities for motility are normally held in check by interactions mediated by cell contact. This notion gained support from observations by Abercrombie that pseudopodial activity was suppressed at sites of cell-cell contact in cultured tissue cells (Fig. 3; Abercrombie, 1970). Inhibition appears to be dependent on the ability to establish *adhaerens*-type cell junctions (Heaysman and Pegrum, 1973).

These ideas suggest the hypothesis that a defective display of contact inhibition of motility in collisions between an invasive cell and cells of the host tissue is necessary for the display of invasive behavior. The hypothesis is supported by the observation that some sarcoma cells do not show contact inhibition following collision with normal fibroblasts (Abercrombie, 1975; Vesely and Weiss, 1973). Although some transformed fibroblasts do display contact inhibition of pseudopodial activity (Bell, 1978; Vasiliev and Gelfand, 1976), the relevance of these observations to the hypothesis in question is uncertain since the invasive behavior of these cells was not determined.

Neutrophil leukocytes showed behavior consistent with the hypothesis (Armstrong and Lackie, 1975). In monolayer culture, neutrophils showed no sign of contact inhibition of pseudopodial activity following collision with fibroblasts. They readily overlapped fibroblasts, moving bodily from the culture substrate onto the dorsal surface of the fibroblast (Fig. 4). Overlapping behavior was never observed following fibroblast-fibroblast collisions.

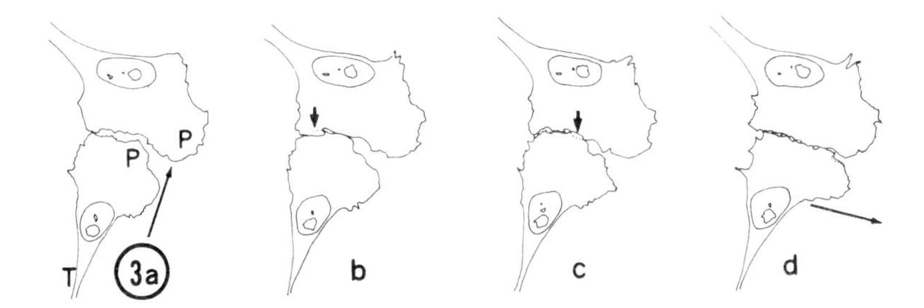

Fig. 3. Contact paralysis of pseudopodial activity following collision of two chick heart fibroblasts in monolayer tissue culture is shown in this series of tracings of selected frames of a time-lapse film. The cultured fibroblast is roughly triangular in shape, with the apex of the triangle being the tail (T) and the leading pseudopod (P) occupying the curved scalloped portion of the cell's margin. The arrow in the first frame (a) indicates the initial direction of motility of the lower cell. Following collision, the pseudopods of both cells are inhibited from further expansion along the margin of contact, limiting overlapping (arrows in b and c). This results in a change in direction of movement of the lower cell, indicated by the arrow in the last frame (d). It can be suggested that paralysis of pseudopods at sites of cell contact reduces cell movement in those coherent tissues in which cells are in contact with other cells on all sides.

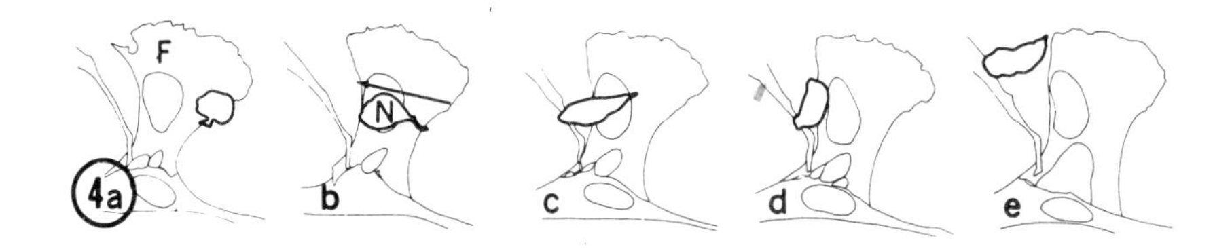

Fig. 4. Tracings of selected frames of a time-lapse movie showing overlapping of a heart fibroblast (F) in monolayer tissue culture by a neutrophil leukocyte (N). The neutrophil moves completely across the dorsal surface of the fibroblasts (b-d, arrow shows the direction of movement).

SUMMARY

A variety of in vitro and in vivo model systems have been employed to analyze the invasive behavior of neutrophil leukocytes. Based on these studies, it appears that tissue growth, destruction of the host tissue, defective tissue recognition, release of serine proteases, and an ability to respond to chemotactic signals are not required for the display of invasive behavior. Factors that may facilitate invasion include low adhesiveness, active motility, high deformability, and an absence of contact paralysis of pseudopodial activity following collision with host tissue cells. It is reasonable to expect that other types of invasive cells may display a different spectrum of processes important in invasion and it is also to be expected that a given cell type might employ different mechanisms when invading different host tissues. Evaluation of invasive mechanisms will, thus, probably require analysis on a cell-type by cell-type basis.

ACKNOWLEDGEMENT

These studies have been supported by NSF Grants #PCM77-18950 and PCM78-18047 and funds from the Cancer Research Coordinating Committee of the University of California. I thank Don Valencia and Edward Wilcox for access to unpublished data. Figure 1 is from Wilhelm Roux' Arch. (1971) 168, 125, © Springer-Verlag, Fig. 2 from Bioscience (1977) 27, 803 © AIBS, and Figs. 3 and 4 from J. Cell Biol. (1975) 65, 439, © Rockefeller University Press.

REFERENCES

Abercrombie, M. (1970). Contact inhibition in tissue culture. In vitro 6, 128-142.

Abercrombie, M. (1973). Ciba Foundation Symposium on Locomotion of Tissue Cells. Elsevier, Amsterdam.

Abercrombie, M. (1975). The contact behavior of invading cells. In Cellular Membranes and Tumor Cell Behavior. Williams and Wilkins, Baltimore, Maryland. pp. 21-37.

Allison, F., M. R. Smith, and W. B. Wood (1955). Studies on the pathogenesis of acute inflammation I. The inflammatory reaction to thermal injury as observed in the rabbit ear chamber. J. Exp. Med., 102, 655-668.

Archer, R. K. (1970). Regulatory mechanisms in eosinophil leukocyte production, release and distribution. In A. S. Gordon (Ed.), Regulation of Hematopoiesis, Vol. II. Appleton-Century-Crofts, New York. pp. 917-941.

Armstrong, P. B. (1966). On the role of metal cations in cellular adhesion: effect on cell surface change. J. Exp. Zool., 163, 99-110.

Armstrong, P. B. (1977). Cellular positional stability and intercellular invasion. Biol. Sci., 27, 803-809.

Armstrong, P. B. (1979). Time-lapse microcinematographic studies of cell motility during morphogenesis of the embryonic yolk sac of Fundulus heteroclitus (Pisces:Teleosti). Submitted for publication.

Armstrong, M. T., and P. B. Armstrong (1978). Cell motility in fibroblast aggregates. J. Cell Sci., 33, 37-52.

Armstrong, M. T., and P. B. Armstrong (1979). The effects of antimicrotubule agents on cell motility in fibroblast aggregates. Exp. Cell Res. In press.

Armstrong, P. B., and J. Lackie (1975). Studies on intercellular invasion in vitro using rabbit peritoneal neutrophil granulocytes (PMN's) I. Role of contact inhibition of locomotion. J. Cell Biol., 65, 439-462.

Ashworth, C. T., V. A. Stembridge, and F. J. Luibell (1961). A study of basement membranes of normal epithelium, carcinoma in situ and invasive carcinoma of uterine cervix utilizing electron microscopy and histochemical methods. Acta Cytol., 5, 369-384.

Atherton, A., and G. V. R. Born (1972). Quantitative investigations of the adhesiveness of circulating polymorphonuclear leukocytes to blood vessel walls. J. Physiol., London, 222, 447-474.
Babai, F. (1976). Etude ultrastructurale sur la pathogénie de l'invasion du muscle strié par des tumenors transplantables. J. Ultrastruct. Res., 56, 287-303.
Babai, F., and G. Tremblay (1972). Ultrastructural study of liver invasion by Novikoff hepatoma. Cancer Res., 32, 2765-2770.
Baggiolini, M. (1972). The enzymes of the granules of polymorphonuclear leukocytes and their functions. Enzyme, 13, 132-160.
Bell, P. B. (1978). Contact inhibition of movement in transformed and nontransformed cells. In R. A. Lerner and D. Bergsma (Eds.), The Molecular Basis of Cell-Cell Interaction (Birth Defects: Original Article Series), Vol. 14, #2, A. R. Liss, New York. pp. 177-194.
Billingham, R. E., and W. K. Silvers (1970). Studies on the migratory behavior of melanocytes in guinea pig skin. J. Exp. Med., 131, 101-117.
Billings-Gagliardi, S. (1977). Mode of locomotion of Schwann cells migrating in vivo. Amer. J. Anat., 150, 73-88.
Blandau, R. J. (1961). Biology of eggs and implantation. In W. C. Young (Ed.), Sex and Internal Secretions, Vol. 2, 3rd ed. Williams and Wilkins. Chap. 14, pp. 797-882.
Blandau, R. J., B. J. White, and R. E. Rumery (1963). Observations on the movements of the living primordial germ cells in the mouse. Fertil. Steril., 14, 482-489.
Boyd, J. D., and W. J. Hamilton (1952). Cleavage, early development and implantation of the egg. In A. S. Parker (Ed.), Marshall's Physiology of Reproduction, Vol. 2. Longman's, London. Chap. 14, pp. 1-126.
Buck, R. C. (1973). Walker 256 tumor implantation in normal and injured peritoneum studied by electron microscopy, scanning electron microscopy, and autoradiography. Cancer Res., 33, 3181-3188.
Butterworth, S. T. G. (1970). Changes in liver lysosomes and cell junctions close to an invasive tumor. J. Pathol., 101, 227-232.
Carr, I., F. McGinty, and P. Norris (1976). The fine structure of neoplastic invasion: invasion of liver, skeletal muscle and lymphatic vessels by the Rd/3 tumour. J. Pathol., 118, 91-99.
Cassiman, J. J., and M. R. Bernfield (1976a). Use of preformed cell aggregates and lazers to measure tissue-specific differences in intercellular adhesions. Develop. Biol., 52, 231-245.
Cassiman, J. J., and M. R. Bernfield (1976b). Transformation-induced alterations in adhesion. Binding of pre-formed cell aggregates to cell layers. Exp. Cell Res., 103, 311-320.
Chew, E. C., R. L. Josephson, and A. C. Wallace (1976). Morphologic aspects of the arrest of circulating cancer cells. In L. Weiss (Ed.), Fundamental Aspects of Metastasis. North Holland Publ., Amsterdam. pp. 121-150.
Clark, E. R. (1912). Further observations on living growing lymphatics: Their relation to the mesenchyme cells. Amer. J. Anat., 13, 351-379.
Clark, E. R., E. L. Clark, and R. O. Rex (1936). Observations on polymorphonuclear leukocytes in the living animal. Amer. J. Anat., 59, 123-173.
Cliff, W. J. (1966). The acute inflammatory reaction in the rabbit ear chamber with particular reference to the phenomenon of leukocytic emigration. J. Exp. Med., 124, 543-556.
Coman, D. R. (1944). Decreased mutual adhesiveness, a property of cells from squamous cell carcinomas. Cancer Res., 4, 625-629.
Coman, D. R. (1961). Adhesiveness and stickiness: two independent properties of the cell surface. Cancer Res., 21, 1436-1438.
Culbertson, C. G. (1971). The pathogenicity of soil amebas. Ann. Rev. Microbiol., 25, 231-254.

Cursons, R. T. M., and T. J. Brown (1978). Use of cell cultures as an indicator of pathogenicity of free-living amoebae. J. Clin. Pathol., 31, 1-11.
de Ridder, L., M. Mareel, and L. Vakaet (1975). Adhesion of malignant and non-malignant cells to cultured embryonic substrata. Cancer Res., 35, 3164-3171.
de Ridder, L., M. Mareel, and L. Vakaet (1977). Invasion of malignant cells into cultured embryonic substrates. Arch. Geschwulstforsch, 47, 7-27.
Di Pasquale, A., and P. B. Bell (1974). The upper cell surface: Its inability to support active cell movement in culture. J. Cell Biol., 62, 198-214.
Dorsey, J. K., and S. Roth (1973). Adhesive specificity in normal and transformed mouse fibroblasts. Develop. Biol., 33, 249-256.
Dubois, R., and Y. Croisille (1970). Germ-cell line and sexual differentiation in birds. Phil. Trans. Roy. Soc. London B, 259, 73-89.
Easty, G. C. (1966). Invasion by cancer cells. In E. J. Ambrose and F. J. C. Roe (Eds.), The Biology of Cancer. Van Nostrand, London. pp. 78-90.
Eaton, R. D. P., E. Meerovitch, and J. W. Costerton (1970). The functional morphology of pathogenicity in Entamoeba histolytica. Ann. Trop. Med. Parasitol., 64, 299-304.
Eaves, G. (1973). The invasive growth of malignant tumors as a purely mechanical process. J. Pathol., 109, 233-237.
Enders, A. C., and S. Schlafke (1972). Implantation in the ferret: Epithelial penetration. Amer. J. Anat., 133, 291-316.
Fell, H. B. (1939). The origin and developmental mechanics of the avian sternum. Phil. Trans. Roy. Soc. London B, 229, 407-464.
Fidler, I. J. (1975). Mechanisms of cancer invasion and metastasis. In F. F. Becker (Ed.), Cancer, Vol. 4. Plenum Press, New York. pp. 101-131.
Fisher, E. B., and B. Fisher (1961). Electron microscopic, histologic and histochemical features of the Walker carcinoma. Cancer Res. 21, 527-531.
Folkman, J., and R. Cotran (1976). Relation of vascular proliferation to tumor growth. Int. Rev. Exp. Pathol., 16, 207-248.
Gabbiani, G. (1979a). The role of contractile proteins in wound healing and fibrocontractive diseases. Meth. Archiev. Exp. Pathol., 9, 187-206.
Gabbiani, G. (1979b). The cytoskeleton in cancer cells in animals and humans. Meth. Archiev. Exp. Pathol., 9, 231-243.
Gallin, J. G., and P. G. Quie (1978). Leukocyte Chemotaxis. Raven Press, New York.
Coldman, R., T. Pollard, and J. Rosenbaum (1976). Cell Motility. (Cold Spring Harbor Conferences on Cell Proliferation, Vol. 3). Cold Spring Harbor Laboratory, Cold Spring Harbor, New York.
Graham, R. C., R. H. Ebert, O. D. Ratnoff, and J. M. Moses (1965). Pathogenesis of inflammation II. In vivo observations of the inflammatory effects of activated Hageman factor and bradykinin. J. Exp. Med., 121, 807-818.
Grant, L. (1973). The sticking and emigration of white blood cells in inflammation. In B. W. Zweifach, L. Grant, and R. T. McCluskey (Eds.), The Inflammatory Process, Vol. 2, 2nd ed. Academic Press, New York. pp. 205-249.
Grant, L., P. Palmer, and A. G. Sanders (1962). The effect of heparin on the sticking of white cells to endothelium in inflammation. J. Pathol. Bact., 83, 127-133.
Haemmerli, G., and P. Sträuli (1978). Motility of LS222 leukemia cells within the mesentery. Virchows Arch. B Cell Pathol., 29, 167-177.
Hamperl, H. (1967). Early invasive growth as seen in uterine cancer and the role of the basal membrane. In P. Denoix (Ed.), Mechanisms of Invasion in Cancer. Springer-Verlag, New York. pp. 17-25.
Heaysman, J. E., and S. M. Pegrum (1973). Early contacts between fibroblasts. An ultrastructural study. Exp. Cell Res., 78, 71-78.
Hirsch, J. G. (1973). Neutrophil leukocytes. In B. W. Zweifach, L. Grant, and R. T. McCluskey (Eds.), The Inflammatory Process, Vol. 1, 2nd ed. Academic Press, New York. pp. 411-447.
Janoff, A. (1972). Neutrophil proteases in inflammation. Ann. Rev. Med., 23, 177-190.

Johnston, M. C., and M. A. Listgarten (1972). Observations on the migration, interaction, and early differentiation of orofacial tissues. In H. C. Slavkin and L. A. Bavetta (Eds.), Developmental Aspects of Oral Biology. Academic Press, New York. pp. 53-80.

Katsuta, H., T. Takaoka, and Y. Nagai (1968). Interaction in culture between normal and tumor cells of rats. In H. Katsuta (Ed.), Cancer Cells in Culture. University of Tokyo Press, Tokyo. pp. 157-168.

Kefalides, N. A. (1978). Biology and Chemistry of Basement Membranes. Academic Press, New York.

Kirby, D. R. S. (1965). The invasiveness of the trophoblast. In The Early Conceptus, Normal and Abnormal. E. and S. Livingstone, London. pp. 68-73.

Lash, J. (1955). Studies on wound closure in urodeles. J. Exp. Zool., 128, 13-28.

Lefford, F. (1972). Cellular emigration from packed cell layers derived from chick embryo donor of different ages. Exp. Cell Res., 70, 242-246.

Leighton, J. (1967). The Spread of Cancer. Academic Press, New York.

Leighton, J. (1968). Bioassay of cancer in matrix tissue culture systems. In The Proliferation and Spread of Neoplastic Cells. M. D. Anderson Hospital Symposium. pp. 533-553.

Lichtman, M. A. (1970). Cellular deformability during maturation of the myeloblast. New Engl. J. Med., 283, 943-948.

Locker, J., P. J. Goldblatt, and J. Leighton (1970). Ultrastructural features of invasion in chick embryo liver metastasis of Yashida ascites hepatoma. Cancer Res., 30, 1632-1644.

Luibel, F. J., E. Sanders, and C. T. Ashworth (1960). An electron microscopic study of carcinoma in situ and invasive carcinoma of the cervix uteri. Cancer Res., 20, 357-361.

Manasek, F. J. (1970). Histogenesis of the embryonic myocardium. Amer. J. Cardiol., 25, 149-168.

Mao, P., K. Nakao, and A. Angrist (1966). Human prostatic carcinoma: An electron microscope study. Cancer Res., 26, 955-973.

Marchesi, V. T. (1970). Mechanisms of blood cell migration across blood vessel walls. In A. S. Gordon (Ed.), Regulation of Hematopoiesis, Vol. 2. Appleton-Century, Crofts, New York. pp. 943-958.

Mareel, M. M. K., and M. J. De Brabander (1978). Effect of microtubule inhibitors on malignant invasion in vitro. J. Nat. Cancer Inst., 61, 787-792.

McGuire, E. J., and C. L. Burdick (1976). Intercellular adhesive selectivity I. An improved assay for the measurement of embryonic cell intercellular adhesion (liver and other tissues). J. Cell Biol., 68, 80-89.

Moyer, W. A., and M. S. Steinberg (1976). Do rates of intercellular adhesion measure the cell affinities reflected in cell sorting and tissue-spreading configurations? Develop. Biol., 52, 246-262.

Nicolson, G. L., and J. L. Winkelhake (1975). Organ specificity of blood-borne tumor metastasis determined by cell adhesion. Nature, London, 255, 230-232.

Nilsson, O. (1962). Electron Microscopy of the human endometrial carcinoma. Cancer Res., 22, 492-494.

Norden, D. M. (1973). The migrating behavior of neural crest cells. In J. F. Bosma (Ed.), Fourth Symp. on Oral Sensation and Perception. U. S. Government Printing Office, Washington, D.C. pp. 9-36.

Ossowski, L., J. P. Quigley, G. M. Kellerman, and E. Reich (1973). Fibrinolysis associated with oncogenic transformation. Requirement of plasminogen for correlated changes in cellular morphology, colony formation in agar, and cell migration. J. Exp. Med., 138, 1056-1064.

Ozzello, L., and P. Sanpitak (1970). Epithelial-stromal junction of introductal carcinoma of the breast. Cancer, 26, 1186-1198.

Patten, B. M., T. C. Kramer, and A. Barry (1948). Valvular action in the embryonic chick heart by localized apposition of endocardial masses. Anat. Rec., 102, 299-311.

Radice, G. P. (1979). An analysis of the mechanism of epidermal wound closure in Xenopus laevae. Doctoral dissertation, Yale University.

Reich, E. (1978). Activation of plasminogen: A widespread mechanism for generating localized extracellular proteolysis. In R. W. Ruddon (Ed.), Biological Markers of Neoplasia: Basic and Applied Aspects. Elsevier, New York. pp. 491-500.

Rifkind, R. A., D. Chin, and H. Epler (1969). An ultrastructural study of early morphogenetic events during the establishment of fetal hepatic erythroporesis. J. Cell Biol., 40, 343-365.

Riley, P. A., and P. Seal (1968). Micro-invasion of epidermis caused by substituted anisoles. Nature, London, 220, 922-923.

Roberson, M. M., and P. B. Armstrong (1979). Regional segregation of Con A receptors on dissociated amphibian embryo cells. Exp. Cell Res. In press.

Romualdez, A. G., and P. A. Ward (1975). A unique complement derived chemotactic factor for tumor cells. Proc. Nat. Acad. Sci., USA, 72, 4128-4132.

Roth, S., E. J. McGuire, and S. Roseman (1971). An assay for intercellular adhesive specificity. J. Cell Biol., 51, 525-535.

Saunders, J. W. (1966). Death in embryonic systems. Science, 154, 604-612.

Schlafke, S., and A. C. Enders (1975). Cellular basis of interaction between trophoblast and uterus at implantation. Biol. Reprod., 12, 41-65.

Schoefl, G. I. (1964). Electron microscopic observations on the regeneration of blood vessels after injury. Ann. N. Y. Acad. Sci., 116, 789-802.

Schoefl, G. I. (1972). The migration of lymphocytes across the vascular endothelium in lymphoid tissue. A reexamination. J. Exp. Med., 136, 568-588.

Seal, P., P. A. Riley, and D. R. Inman (1969). Basal cell encroachment into the dermis caused by application of hydroxyanisole. J. Invest. Dermatol., 52, 264-267.

Sherman, M. I., and L. R. Wudl (1976). The implanting mouse blastocyst. In G. Poste and G. L. Nicolson (Eds.), The Cell Surface in Animal Embryonesis and Development. Elsevier/North-Holland, Amsterdam. pp. 81-125.

Speidel, C. C. (1964). In vitro studies of myelinated nerve fibers. Int. Rev. Cytol., 16, 173.

Speirs, R. S. (1970). Function of leukocyte in inflammation and immunity. In A. S. Gordon (Ed.), Regulation of Hematopoiesis, Vol. 2. Appleton-Century-Croft, New York. pp. 995-1043.

Speirs, R. S., and Y. Osada (1962). Chemotactic activity and phagocytosis of eosinophils. Proc. Soc. Exp. Biol. Med., 109, 929-932.

Steinberg, M. S. (1970). Does differential adhesion govern self-assembly processes in histogenesis? Equilibrium configurations and the emergence of a hierarchy among populations of embryonic cells. J. Exp. Zool., 173, 395-434.

Steinberg, M. S., P. B. Armstrong, and R. E. Granger (1973). On the recovery of adhesiveness by trypsin-dissociated cells. J. Membrane Biol., 13, 97-128.

Sträuli, P., and L. Weiss (1977). Cell locomotion and tumor penetration. Europ. J. Cancer, 13, 1-12.

Strickland, S., E. Reich, and M. I. Sherman (1976). Plasminogen activator in early embryogenesis: Enzyme production by trophoblast and parietal endoderm. Cell, 9, 231-240.

Sugar, J. (1968). An electron microscopic study of early invasive growth in human skin tumors and laryngeal carcinoma. Europ. J. Cancer, 4, 33-38.

Takeichi, M., H. S. Ozaki, K. Tokunaga, and T. S. Okada (1979). Experimental manipulation of cell surface to affect cellular recognition mechanisms. Devel. Biol., 70, 195-205.

Tarin, D. (1967). Sequential electron microscopical study of experimental mouse skin carcinogenesis. Int. J. Cancer, 2, 195-211.

Tickle, C., A. Crawley, and M. Goodman (1978a). Cell movement and the mechanism of invasiveness: A survey of the behavior of some normal and malignant cells implanted into the developing chick wing bud. J. Cell Sci., 31, 293-322.

Tickle, C., A. Crawley, and M. Goodman (1978b). Mechanisms of invasiveness of epithelial tumors: Ultrastructure of the interactions of carcinoma cells with embryonic mesenchyme and epithelium. J. Cell Sci., 33, 133-155.

Twitty, V. C. (1949). Developmental analysis of amphibian pigmentation. In Growth. Ninth Symp. on Devel. and Growth, 13, suppl., 133-161.

Twitty, V. C., and M. C. Niu (1954). The motivation of cell migration, studied by isolation of embryonic pigment cells singly and in small groups in vitro. J. Exp. Zool., 125, 541-573.

Vasiliev, J. M., and I. M. Gelfand (1976). Morphogenetic reactions and locomotory behavior of transformed cells in culture. In L. Weiss (Ed.), Fundamental Aspects of Metastasis. North-Holland, Amsterdam. pp. 71-98.

Vassalli, J. D., A. Gravelli-Piperno, C. Griscelli, and E. Reich (1978). Specific protease deficiency in polymorphonuclear leukocytes of Chédiak-Higashi syndrome and beige mice. J. Exp. Med., 147, 1285-1290.

Veselý, P., and R. A. Weiss (1973). Cell locomotion and contact inhibition of normal and neoplastic rat cells. Int. J. Cancer, 11, 64-76.

Vracko, R. (1974). Basal lamina scaffold-anatomy and significance for maintenance of orderly tissue structure. Amer. J. Path., 77, 314-346.

Walther, B. T., R. Öhman, and S. Roseman (1973). A quantitative assay for intercellular adhesion. Proc. Nat. Acad. Sci. USA, 70, 1569-1573.

Weinstein, R. S., F. B. Merk, and J. Alroy (1976). The structure and function of intercellular junctions in cancer. Adv. Cancer Res., 23, 23-89.

Weiss, L. (1977). Cell detachment and metastasis. In P. G. Stansly and H. Sato (Eds.), Cancer Metastasis (Gann Monogr. Cancer Res. 20). University of Tokyo Press, Tokyo. pp. 25-35.

Weiss, L., and R. R. A. Coombs (1963). The demonstration of rupture of cell surfaces by an immunological technique. Exp. Cell Res., 30, 331-338.

Wilkinson, P. C. (1974). Chemotaxis and Inflammation. Churchill, Livingstone, Edinburgh.

Wilkinson, P. C. (1978). The adhesion, locomotion, and chemotaxis of leukocytes. In J. R. Vane and S. H. Ferreira (Eds.), Handbook of Experimental Pharmacology, Vol. 50/1 (Inflammation). Springer-Verlag, Berlin. pp. 109-137.

Willis, R. A. (1960). Pathology of Tumors, 3rd ed. Butterworths, London.

Wilson, I. B., and D. M. Potts (1970). Melanoma invasion in the mouse uterus. J. Reprod. Fert., 22, 429-434.

Witschi, E. (1948). Migration of the germ cells of human embryos from the yolk sac to the primitive gonadial folds. Carnegie Inst. Wash. Contr. Embryol. 32, 67-80.

Wolff, E. (1963). Long-term organotypic culture of human surgical tumors at the expense of substances elaborated by the mesonephros of the chick embryo. Nat. Cancer Inst. Monogr., 11, 180-195.

Wolff, E., and E. Wolff (1967). Factors of growth and maintenance of tumors as organized structures in vitro. Cell Differentiation, Ciba Foundation Symposium. Little, Brown and Co., Boston. pp. 208-218.

Wood, S. (1958). Pathogenesis of metastasis formation observed in vivo in the rabbit ear chamber. A. M. A. Arch. Pathol. 66, 550-568.

Wood, S., R. R. Baker, and B. Marzocchi (1957). Locomotion of cancer cells in vivo compared with normal cells. In P. Denoix (Ed.), Mechanisms of Invasion in Cancer (UICC Monogr. #6). Springer-Verlag, Berlin. pp. 26-30.

Wood, S., R. R. Baker, and B. Marzocchi (1968). In vivo studies of tumor behavior: Locomotion of and interrelationships between normal cells and cancer cells. In The Proliferation and Spread of Neoplastic Cells (M. D. Anderson Hospital Symposium). Williams and Wilkins, Baltimore. pp. 495-509.

Woods, D. A., and C. J. Smith (1969). Ultrastructure of the dermal-epidermal junction in experimentally induced tumors and human oral lesions. J. Invest. Dermatol., 52, 259-263.

Young, J. S. (1959). The invasive growth of malignant tumors: An experimental interpretation based on elastic-jelly models. J. Pathol. Bacteriol., 77, 321-339.
Zigmond, S. H. (1978). Chemotaxis by polymorphonuclear leukocytes. J. Cell Biol., 77, 269-287.

Cell Movement and Invasion *in vivo*

Embryo Implantation and Trophoblast Invasion

H. -W. Denker

Abteilung Anatomie der RWTH, Melatener Str. 211, D-5100 Aachen, Federal Republic of Germany

ABSTRACT

The trophoblast of the implanting mammalian embryo receives interest for being a highly invasive non-tumor tissue. This becomes particularly obvious when trophoblast is transplanted to ectopic sites.

The physiological sequence of events during implantation of the embryo in the uterus involves apposition, dissolution of extracellular blastocyst coverings, physicochemical changes in the cell surface coats of trophoblast and uterine epithelium, adhesion, and, in most species, penetration of the trophoblast through the uterine epithelium towards subepithelial blood vessels. Various modes of penetration (displacement, cell fusion, intrusion) are described. Many observations suggest that the relation between invasiveness of the trophoblast and readiness of the uterine epithelium to degenerate varies from one species to the other.

The mechanism of the interaction between trophoblast and uterine epithelium which leads to implantation initiation is still largely unknown in spite of intensive research efforts. Recently evidence has been found for an important role played by certain proteinases of these tissues. In the rabbit, a peculiar proteinase of the implanting trophoblast (called "blastolemmase") seems to be essential for implantation initiation. Preliminary biochemical characterization has been achieved. Specific proteinase inhibitors interfere with implantation when administered in vivo. The possible role of this proteinase and of related enzymes in attachment and/or invasion is discussed.

KEYWORDS

Embryo implantation; trophoblast; adhesion; epithelial penetration; invasion; glycoproteins; blastolemmase; control by inhibitors.

INTRODUCTION

Implantation is the process by which an intimate cellular contact is formed between embryonic and maternal tissues, as typically found in eutherian mammals. In histological terms, the involved partners are: 1. the <u>trophoblast</u>, i.e. a specific extraembryonic population of cells of the conceptus which appears to be specialized for the formation of this contact (and, besides this, for nutrition and hormone production), and 2. the endometrium of the uterus. In most species including the human, implantation involves true invasion of the trophoblast into the endometrium. The cytological details of this process are of interest because they resemble, in various respects, tumor invasion. The destructive potential of the trophoblast becomes particularly obvious when these cells are transplanted to ectopic sites, as described below. However, in the hormonally conditioned uterus is trophoblastic invasion limited in time and space and is halted before causing total destruction of the host organ. Instead, a complex morphogenetic process which involves both partner tissues is initiated at the border between them, leading to the formation of a highly specialized exchange organ, the placenta.

We will consider here only those species in which a truly invasive placenta is formed, omitting the epitheliochorial placentation found e.g. in ungulates where trophoblast and uterine epithelium are only tightly apposed against (and adhere to) each other. Interestingly, it has been reported that even in these latter species shows the trophoblast invasive properties if transplanted to ectopic sites (Samuel, 1971).

CYTOLOGICAL DETAILS OF THE INTERACTION BETWEEN TROPHOBLAST AND ENDOMETRIUM

Recently attention of investigators is focused on the initiation phase of implantation. One of the most astonishing phenomena in this process is its initiation by formation of a cellular contact between the apical parts of cytoplasm of the two epithelia , the trophoblast and the uterine epithelium. Implantation then continues by penetration of the trophoblast through the uterine epithelium followed by invasion into the deeper parts of the endometrium. The cytological details to be described below are found quite consistently and seem to be largely independent of the type of topographical relationship between the blastocyst and the uterus present in the particular species at the time of implantation, i.e. the centric (large blastocyst which fills the uterine lumen),the eccentric (small blastocyst which implants in an endometrial crypt which it fills first)or the interstitial type (small blastocyst which penetrates, as a whole, through a minor defect in the uterine epithelium). In particular, there is no consistent correlation between these three types of topographical relationship and the three modes of epithelial penetration to be described below.

Electron microscopical studies performed in a number of species during the last years have shown that there is a certain rule in the sequence of cytological details seen during the initial phase of implantation (for review, see Schlafke and Enders, 1975). In the first phase, a stable cellular contact is formed between the blastocyst, which was before freely movable in the uterine lumen, and a portion of the uterine epithelium. This phase consists of two stages, apposition and adhesion.

During apposition, the blastocyst becomes immobilized and those parts of the endometrium which will enter into interaction with the trophoblast are determined. The mechanism of this process and the morphological details vary considerably from one species to the other. In species with small blastocysts (rodents), a local oedema induced in the endometrium in the vicinity of the blastocyst is involved in the immobilization of the embryo, whereas myometrial contractions play a significant role in species with large blastocysts (rabbit, see Böving, 1963).

At this stage, the blastocysts are still surrounded in a number of species by thick coats of extracellular material, the so-called blastocyst coverings, i.e. the zona pellucida or more complex, multilayered structures (as seen in the rabbit and the fur seal) (Böving, 1963; Denker, 1970, 1977; Denker and Gerdes, 1979). There is evidence that physicochemical changes which these coverings undergo in this phase play a role in the process of apposition: the adhesiveness of the coverings increases considerably (especially at the abembryonic pole, i.e. where implantation begins, as shown in the rabbit). It has been postulated that this local change in adhesiveness is involved in establishing the correct orientation of the blastocyst in the uterus, with its abembryonic (trophoblastic) pole facing the antimesometrial endometrium (Böving, 1963; Denker, 1978). Enzyme activities probably involved in this process will be discussed below.

The increase in adhesiveness is directly followed by dissolution of the blastocyst coverings, and the trophoblast establishes a cellular contact with the uterine epithelium immediately thereafter, in the rabbit. This close sequence of events has found interest because there might be a connection between the causal mechanisms. The same is possibly valid also for other species (e.g. the ferret) including also the murine rodents (mouse, rat etc.) although there has been much confusion in the literature, because in the latter group the blastocysts are able to hatch mechanically from their zona pellucida and to remain unimplanted in the uterine lumen, as seen during lactation-induced delay of implantation. However in regular pregnancy, the described sequence of events is found quite typically (for references, see Denker, 1977). In several other species (cat, ungulates) the dissolution of blastocyst coverings and formation of a cellular contact between trophoblast and endometrium are always clearly separated in time. The situation in the human remains uncertain because specimens from the apposition and attachment phases are lacking, and the investigated unattached human blastocysts have been fixed with solutions which are unsuitable for preservation of the coverings. In the guinea pig which shows an interstitial implantation like human embryos, ectoplasmic processes of trophoblast cells penetrate through the zona which is apparently dissolved by proteolytic enzymes; immediately afterwards they establish contact with the uterine epithelium (Spee, 1901; Blandau, 1949; Parr, 1973).

After the blastocyst coverings have been disposed of, the trophoblast approaches the surface of the uterine epithelium, and the microvilli of both epithelia interdigitate showing a more or less regular pattern, depending on the species. Even at this stage of apposition, the blastocysts can still be flushed out of the uterus using slight pressure, although a certain degree of adhesion has doubtless been reached.

The morphologically observable intimacy of the contact between both epithelia is thereafter gradually increased. In terms of electron microscopical classification the adhesion stage is reached when the apical cell membranes of trophoblast and uterine epithelium not only approach each other focally but run parallel to one another over longer distances, and there are regions of very close (less than 200 Å) membrane association. The microvilli of both epithelia which had before shown regular interdigitation as described above flatten, and the parallel running cell membranes form an irregular, waved contour (Enders and Schlafke, 1967; Reinius, 1967; Bergström, 1971; Parkening, 1976). This sequence of events is found more or less typically in all species investigated so far. In some species, however, only certain parts of the trophoblast establish the contact mentioned. In the rabbit, for example, the initial contact is formed only by specialized elements of the trophoblast, the syncytial "trophoblastic knobs" (Böving, 1963; Denker, 1970, 1977). In the murine rodents, the adhesion takes place fairly uniformly over the entire trophoblast surface. The two sets of epithelial cells which have established contact in the described way show little cytological specialization except for prominent ectoplasmic regions which are often rich in microfilaments.

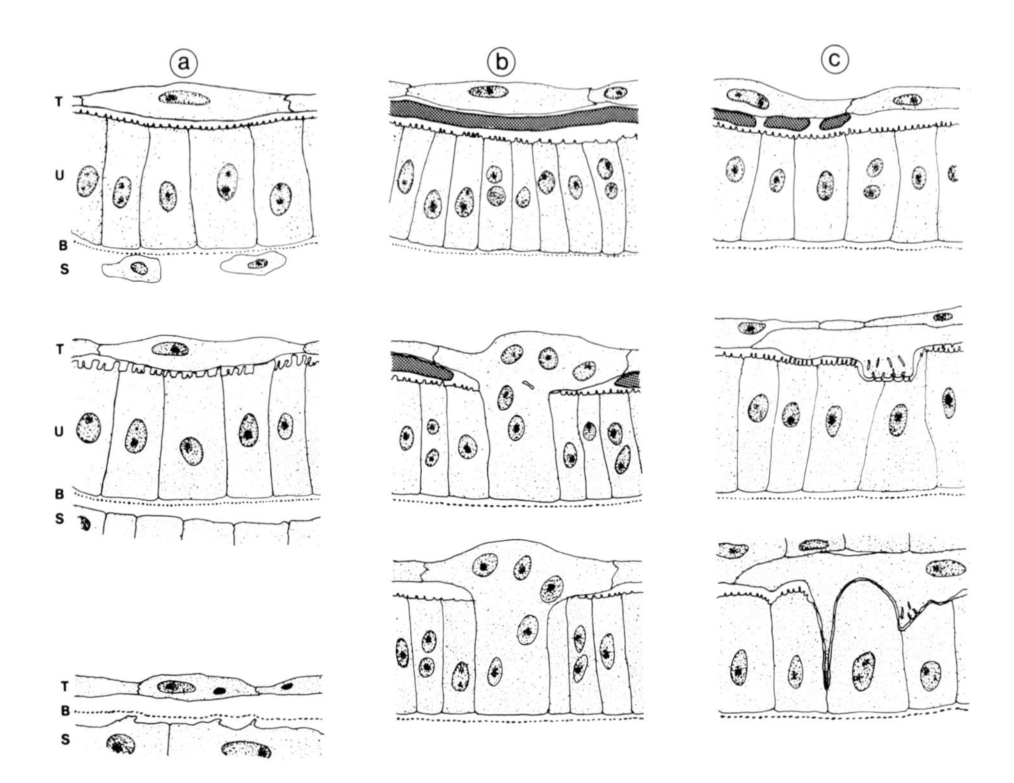

Fig. 1. The three types of interaction of trophoblast with uterine epithelium during the apposition, adhesion and penetration phases of implantation. (a) Penetration by displacement (rat, mouse); (b) penetration by fusion (rabbit); (c) penetration by intrusion (ferret, other species?). (Adopted, in modified form, from Schlafke and Enders, 1975, by permission of the authors and the editor, Biology of Reproduction)
T: trophoblast; U: uterine epithelium; B: basal lamina;
S: stroma cells being transformed into decidual cells in (a).

The cellular contact is reinforced in the beginning of the next phase, epithelial penetration. Typical junctional complexes including even desmosomes are formed between the invading trophoblast and the uterine epithelium. This phenomenon has received interest since it is known that, apart from formation of mechanical contact, certain junctions can mediate ionic coupling and information transfer which seems remarkable when it occurs between two epithelia of different organisms, the embryo and the mother.

A comparative analysis of the cytological details observed in various species has shown that three different types of epithelial penetration can be distinguished as illustrated diagrammatically in Fig. 1 (Schlafke and Enders, 1975):

1. Penetration by displacement: In a number of species (like murine rodents), the uterine epithelium is sloughed off the basal lamina in relatively large areas adjacent to the implanting blastocyst. By this way the trophoblast can come into contact with the denuded basal lamina, and it then grows between the latter and neighboring epithelial cells thus apparently contributing to the process of displacement. Sloughed epithelial cells are found both in groups and as individual cells, and are phagocytized by the trophoblast. Numerous investigations have well documented that the uterine epithelium has, in these species, a pronounced tendency to undergo the described sloughing, even after nonspecific mechanical stimuli. There is some evidence that this tendency may be a peculiar property of the uterine epithelial cells themselves, preprogrammed under proper hormonal stimulation. It may require specific genetic activity because the epithelium remains intact after actinomycin D treatment (Finn and Bredl, 1973). There is also evidence that the loss of the epithelium partly depends in addition on isolation from the blood supply by differentiation of the decidual cells which form a "barrier" under the basal lamina (Fig. 1, a). Formation of decidual cells is in fact induced by the same kind of stimuli (blastocyst, mechanical alteration) mentioned above. Interestingly, the actinomycin experiments have shown that the trophoblast does have the ability to penetrate into epithelium which appears to be intact, even in these species. Displacement penetration, therefore, may be a very special situation, and regular implantation in the uterus as seen in these species may not be a very suitable model for studies of trophoblast invasiveness.

2. Penetration by fusion: Initiation of implantation by fusion of trophoblast with uterine luminal epithelium has been well documented electron microscopically in the rabbit (Enders and Schlafke, 1971). Material from primates and the human as investigated so far leaves some doubt whether fusion might also take place in these species (Böving and Larsen, 1973; Schlafke and Enders, 1975). In the rabbit, syncytial elements of the trophoblast, the so-called trophoblastic knobs, penetrate through the extracellular blastocyst coverings, adhere (as described above) to the surface of uterine epithelial cells which are overlying a subepithelial capillary (Böving, 1963), and, immediately thereafter, fuse with them so that a compound symplasm is formed. The fate of the maternal nuclei in this symplasm is not known exactly, although there is no indication of degeneration. In any case, the fact that the number of trophoblast nuclei in this early fusion stage is larger than that of uterine nuclei appears to be sufficient to induce a change in cellular behavior so that the basal lamina will be penetrated and the subepithelial capillary be arroded until a hemochorial contact is formed (see Denker, 1977: Fig. 4). The trophoblast is particularly rich in microfilaments, and it should be interesting to know more about the role of these structures and of microtubules in the process of invasion. The description just given is based on the details seen during formation of the yolk sac placenta, but basically the same phenomena are observed at the formation of the chorioallantoic placenta which follows, with the only difference that both the uterine cavum epithelium and the trophoblast are first transformed into broad symplasms, which then show extensive fusion (Larsen, 1961).

3. Penetration by intrusion: This type of epithelial penetration has been documented best in the ferret (Enders and Schlafke, 1972). Histological evidence from other species which have been studied only or predominantly with the light microscope suggests that intrusion penetration may be found in many species (including the guinea pig and possibly primates). In the ferret, the trophoblast of the implanting blastocyst first forms syncytial elements which are comparable to the trophoblastic knobs seen in the rabbit. Ectoplasmic pads of syncytial trophoblast attach to the surface of the uterine epithelium and often indent it. This is followed by penetration of thin processes of the trophoblast between the apical ends of uterine peithelial cells. Interestingly, these parts of the trophoblast then share the apical junctional complexes of the uterine epithelium, and a number

of desmosomes are formed between both partners. Subsequently, the trophoblast processes traverse the basal lamina and rapidly invade the stroma where they surround the blood vessels forming the endotheliochorial contact typical for carnivores. The described trophoblast processes are rich in microfilaments and also possess microtubules.

Trophoblast invasion seems to be halted for a while before penetrating through the basal lamina, in all three types of implantation. The mechanism of overcoming this barrier is unknown.

The following phase of deeper penetration into the endometrial stroma is quite impressive from the histological point of view because the invasive properties of the trophoblast become perhaps even more obvious. The intimacy of contact finally reached is traditionally described using Grosser's classification (endotheliochorial and hemochorial contact being the principal types seen in invasive placentae) although this contributes little to understanding the cell physiological properties of the trophoblast. Purely desciptive morphological studies of implantation as seen in regular pregnancy have generally given only very limited information to those interested in the mechanisms of invasiveness.

ASPECTS OF ECTOPIC TROPHOBLAST GROWTH AND THE ROLE OF HOST TISSUE IN REGULATION OF INVASION

When transplanted to extrauterine (ectopic) sites, the trophoblast shows its invasive potential most impressively. Ectopic growth of the trophoblast has been studied extensively in the mouse where it can easily be obtained after transplanting preimplantation stage embryos to such sites as the kidney or the testis (Kirby, 1970; for review see Billington, 1971). In this situation it becomes particularly obvious that trophoblast invasion is accompanied by destruction and phagocytosis of host tissues, although these phenomena form also part of the regular implantation process in the endometrium. It was shown that transplanted trophoblast can cause destruction even of mouse mammary carcinoma (Kirby, 1962). The type of host tissue seems to determinate, at least to a certain extent, whether cytolysis at distance dominates (as seen in the testis) or whether vital appearing host cells are engulfed and phagocytosed. In the latter case, cell junctions including desmosomes can be formed between trophoblast and host cells (in the same way as in regular uterine implantation, see above) as described in case of human choriocarcinoma growing in hamster liver.

A widely favored hypothesis postulates that trophoblast invasion in the uterus is regulated by the decidua which may function as a barrier, possessing inhibitory properties of unknown nature. In fact, trophoblast transplanted into a non-decidualized uterus (lacking proper hormonal conditioning) in the mouse causes marked destruction of the host organ up to the level of the myometrium. This may be comparable to the situation of placenta accreta in man. It has, therefore, been suggested that, in regular pregnancy, the extent of trophoblast invasion is determined by the limits of an area in which the decidual cells have a tendency to degenerate even spontaneously (for review, see Billington, 1971; Finn, 1971). On the other hand, sarcoma cells transplanted to the hormonally conditioned uterus were found to invade right through the decidua to the myometrium (Smith and Hartman, 1974). Furthermore, the example of ectopic implantation shows that the trophoblast seems to have an inherently limited life-span: the invasive growth of extrauterine trophoblast lasts only a few days longer than the time at which it normally ceases in the uterus, and it finally degenerates at approximately the same time at which pregnancy would be terminated. It may be significant that invasion stops before degeneration of trophoblast cells becomes apparent (Sherman and Wudl, 1976), and this may indicate that proliferation and invasion are not necessarily connected. Anyway the stop of invasion cannot simply be explained by disappearance of cytotrophoblastic "stem" cells as suggested for the situation in the human, because it was shown that cytotrophoblast elements can persist until term. Attempts to establish trophoblast-derived cell lines in long term in vitro culture have met with varying success. A trophoblast tumor with unlimited growth characteristics (the choriocarcinoma) is well known in the human but no adequate correlate was found in any other species (for review, see Hertz, 1978).

THE ROLE OF CERTAIN PROTEINASES IN EMBRYO IMPLANTATION

It has often been assumed that various hydrolytic enzymes may be involved in the process of implantation. In particular, enzymes which attack cell surface glycoproteins and/or intercellular ground substance glycosaminoglycans (i.e. glycosidases and proteases) have been regarded as good candidates (Denker, 1970, 1971 a and b). As a matter of fact, the structures which must be overcome during formation of an intimate cellular contact between trophoblast and endometrium, i.e. the blastocyst coverings (zona pellucida and/or its analogs) and the thick surface coat of the uterine epithelium, were found to be very rich in glycoproteins (with sialic acid and sulfate ester groups) (Denker, 1970). The increased adhesiveness of the blastocyst coverings and of the trophoblast surface as observed at implantation initiation may be due to physicochemical changes in these glycoproteins.

Of all enzymes investigated, only certain proteinases (endopeptidases) have so far been shown to play an indispensable and crucial role in implantation initiation (Denker, 1977). Interest has been focused on a peculiar trophoblast-dependent endopeptidase called blastolemmase studied extensively in the rabbit. This enzyme belongs to the trypsin family but has more narrow substrate specificity than trypsin itself (Denker and Fritz, 1979). Probably in concerted action together with uterine secretion proteinase(s)(and perhaps with glycosidases also shown to be present), blastolemmase appears to play a central role in initiation of implantation. When rabbits are treated intrauterally with specific proteinase inhibitors (which were selected on the basis of blastolemmase inhibition in vitro), the dissolution of the blastocyst coverings is blocked and the trophoblast cannot attach to the uterine epithelium (Fig. 2 and 3) (Denker, 1977). Inhibitors studied were: aprotinin (Trasylol), antipain, boar seminal plasma trypsin-acrosin inhibitor, and p-nitrophenyl-p'-guanidinobenzoate (NPGB). On contrast, epsilon-aminocaproic acid, an inhibitor of plasminogen activation which does not inhibit blastolemmase, is without any effect on this process. While it became clear from the described experiments that blastolemmase plays an indispensable role in the dissolution of blastocyst coverings as part of the processes leading to implantation, it is not certain at the present time whether this enzyme is directly involved in initiation of attachment of the trophoblast to the uterine epithelium and its subsequent fusion and invasion: In inhibitor-treated uteri, occasional attachment and fusion, although restricted to few small areas, was observed in regions where interposed blastocyst coverings were shed due to mechanical rupturing as a result of the continuing expansion of the blastocyst. Experiments presently being performed in this laboratory will hopefully answer this question.

It should be interesting to have comparative data from other species in which far less is known about the possible role of proteinases in implantation. Trypsin-like enzyme(s) seem(s) to be essential for implantation in the mouse (for review see Denker, 1977). In the cat, a cathepsin B-type enzyme predominates, the physiological function of which is unknown (Denker and coworkers, 1978). An enzyme with trypsin-like activity was also found in implanting blastocysts in the guinea pig (Denker, unpublished). The question whether this is related to the gelatinolytic activity found at the same stage in this species (Blandau, 1949) will have to be answered by further experiments.

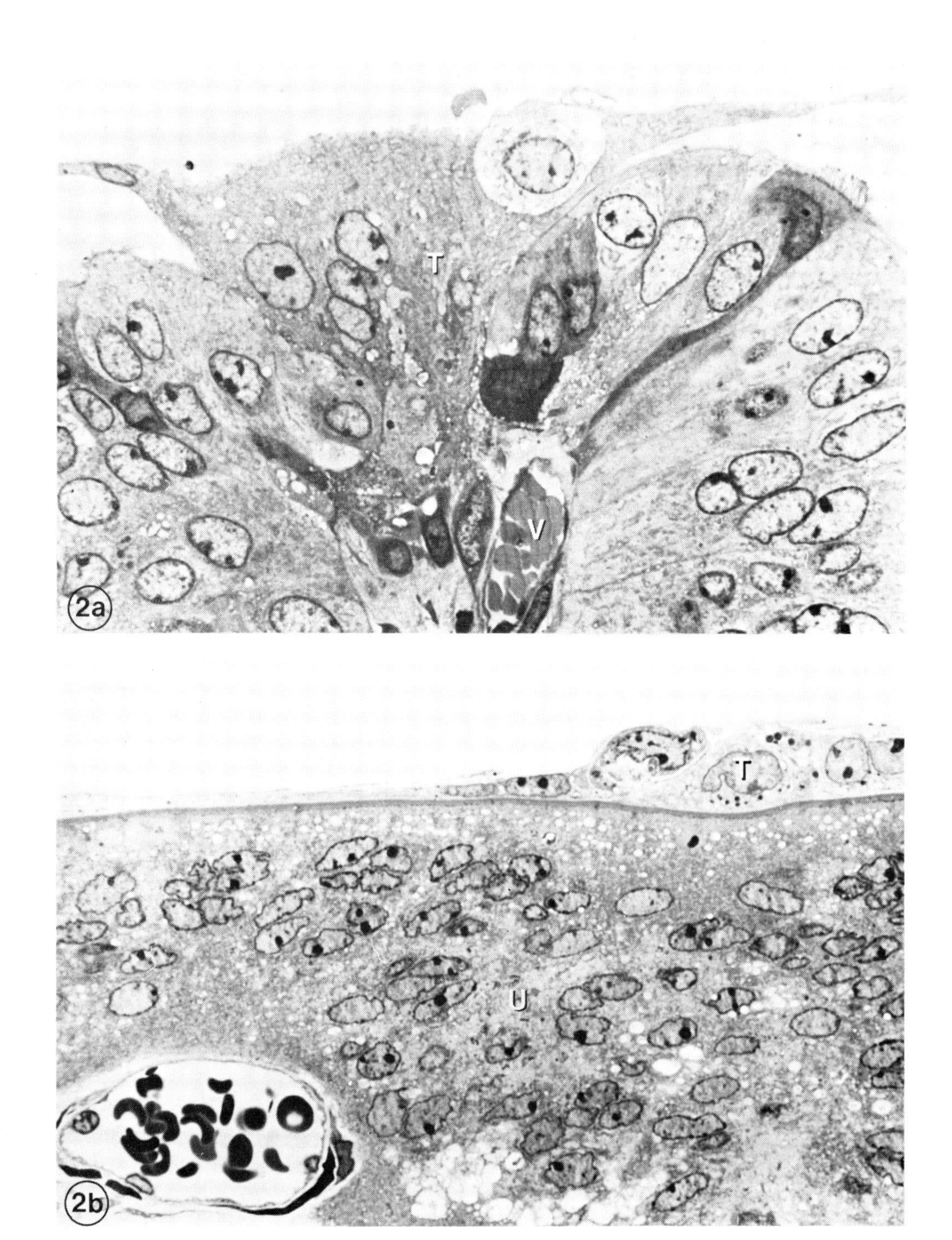
T
V
2a
T
U
2b

In addition to blastolemmase, trypsin family enzymes which have even narrower substrate specificity are present in implanting rabbit blastocysts (Denker and Petzoldt, in preparation). Such enzymes might be involved in changing, by limited proteolysis, cell surface properties (including adhesiveness and receptor functions) relevant to attachment and invasion. This aspect is being studied in recent experiments.

ACKNOWLEDGEMENT

The author wishes to thank cordially Mrs. Gerda Bohr and Edith Höricht for excellent technical assistance, Mrs. Ria Becht for typing the manuscript, Mr. W. Graulich for drawing the diagram,and to Drs. Sandra Schlafke and A.C. Enders for the permission to use their illustration (Fig. 1). Author's own investigations were supported by Deutsche Forschungsgemeinschaft grants No. De 181/3-8.

Fig. 2 (Opposite page) Light micrographs of semithin sections illustrating inhibition of embryo implantation by administration of proteinase inhibitors in the rabbit. x 1000.

a) Control, 7 1/2 d p.c. The syncytial trophoblastic knob (T) has penetrated the uterine epithelium and has already nearly reached the subepithelial blood vessel (V).

b) Proteinase activity was inhibited by administration of 6 mg of aprotinin (Trasylol) into the uterine lumen at 6 1/2 d p.c. The stage shown is 8 1/2 d p.c., i.e. one day later than the control (a). Even at this time, the blastocyst coverings (dark line) are not dissolved yet and are interposed between trophoblast and uterine epithelium (U) (already transformed into a broad symplasm). The trophoblastic knob (T) has not been able to attach, to invade and to reach the subepithelial blood vessel (lower left corner).

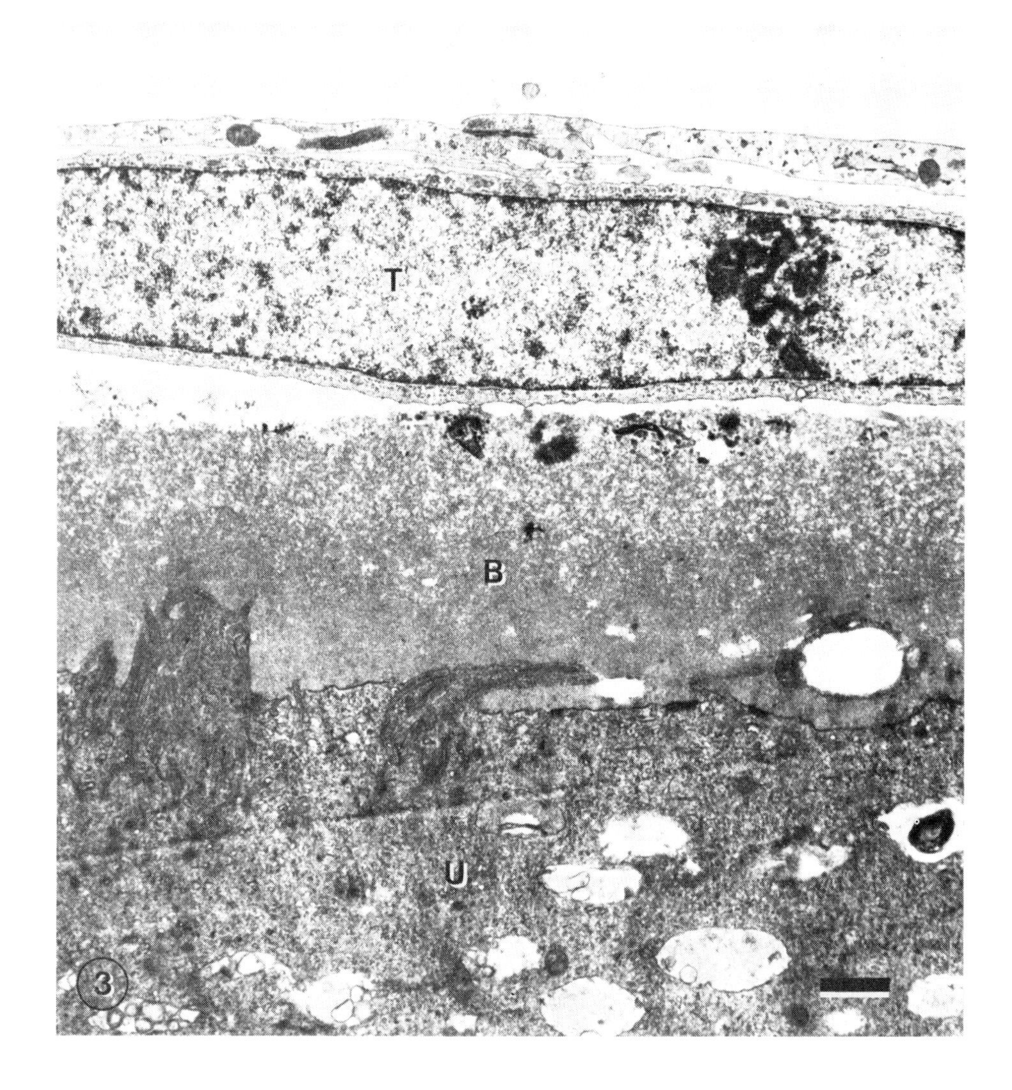

Fig.3. Electron micrograph showing the extracellular glycoprotein material of the blastocyst coverings (B) still interposed between rabbit trophoblast (T) and uterine epithelium (U) at 8 1/2 d p.c. Proteinase activity was inhibited by intrauterine administration of 6 mg of aprotinin (Trasylol) at 6 1/2 d p.c. Even 1 1/2 days after the regular time of implantation, no cellular contact has been established between trophoblast and uterine epithelium. The apical part of the latter has formed hemidesmosome-like structures where it touches the blastocyst coverings. Bar = 1 μm.

REFERENCES

Bergström, S. (1971). Surface Ultrastructure of Mouse Blastocysts Before and at Implantation. Thesis, Uppsala.

Billington, W.D. (1971). Biology of the trophoblast. In: M.W.H.Bishop (Ed.), Advances in Reproductive Physiology, Vol. 5. Academic Press, New York/London.pp.27-66.

Blandau, R.J. (1949). Embryo-endometrial interrelationship in the rat and guinea pig. Anat. Rec., 104, 331-359.

Böving, B.G. (1963). Implantation mechanisms. In: C.G. Hartman (Ed.), Conference on Physiological Mechanisms Concerned with Conception. Pergamon Press, Oxford. pp. 321-396.

Böving, B.G., and J.F. Larsen (1973). Implantation. In: E.S.E.Hafez and T.N. Evans (Eds.), Human Reproduction: Conception and Contraception. Harper & Row, Hagerstown (Maryland). pp. 133-156.

Denker, H.-W. (1970). Topochemie hochmolekularer Kohlenhydratsubstanzen in Frühentwicklung und Implantation des Kaninchens. I and II. Zool. Jahrb., Abt. Allgem. Zool. Physiol., 75, 141-245 and 246-308.

Denker, H.-W. (1971a). Enzym-Topochemie von Frühentwicklung und Implantation des Kaninchens. II. Glykosidasen. Histochemie, 25, 268-285.

Denker, H.-W. (1971b). Enzym-Topochemie von Frühentwicklung und Implantation des Kaninchens. III. Proteasen. Histochemie, 25, 344-360.

Denker, H.-W. (1977). Implantation: The Role of Proteinases, and Blockage of Implantation by Proteinase Inhibitors. Springer-Verlag, Berlin - Heidelberg - New York. (= Adv. Anat. Embryol. Cell Biol. Vol. 53 Fasc. 5).

Denker, H.-W. (1978). The role of trophoblastic factors in implantation. In: Ch.H. Spilman and J.W. Wilks (Eds.), Novel Aspects of Reproductive Physiology. Spectrum Publications (distr. by Halsted Press/John Wiley & Sons), New York. pp. 181-212.

Denker, H.-W., L.A. Eng, and C.E. Hamner (1978). Studies on the early development and implantation in the cat. II. Implantation: proteinases. Anat. Embryol., 154, 39-54.

Denker, H.-W., and H. Fritz (1979). Enzymic characterization of rabbit blastocyst proteinase with synthetic substrates of trypsin-like enzymes. Hoppe-Seyler's Z. Physiol. Chem., 360, 107-113.

Denker, H.-W., and H.-J. Gerdes (1979). The dynamic structure of rabbit blastocyst coverings. I. Transformation during regular preimplantation development. Anat. Embryol. (in press).

Enders, A.C., and S. Schlafke (1967). A morphological analysis of the early implantation stages in the rat. Am.J.Anat., 120, 185-226.

Enders, A.C., and S. Schlafke (1971). Penetration of the uterine epithelium during implantation in the rabbit. Am.J.Anat., 132, 219-240.

Enders, A.C., and S. Schlafke (1972). Implantation in the ferret: Epithelial penetration. Am.J.Anat., 133, 291-316.

Finn, A.C. (1971). The biology of decidual cells. In: M.W.H. Bishop (Ed.),Advances in Reproductive Physiology, Vol.5. Academic Press, New York/London. pp. 1-26.

Finn, C.A., and J.C.S. Bredl (1973). Studies on the development of the implantation reaction in the mouse uterus: influence of actinomycin D. J. Reprod. Fert., 34, 247-253.

Hertz, R. (1978). Choriocarcinoma and Related Gestational Trophoblastic Tumors in Women. Raven Press, New York.

Kirby, D.R.S. (1962). Ability of the trophoblast to destroy cancer tissue. Nature, 194, 696-697.

Kirby, D.R.S. (1970). The extra-uterine mouse egg as an experimental model. In: G.Raspé (Ed.), Schering Symposium on Mechanisms Involved in Conception. Adv. Biosciences, Vol. 4. Pergamon Press/Vieweg, Oxford etc. pp. 255-273.

Larsen, J.F. (1961). Electron microscopy of the implantation site in the rabbit. Am.J.Anat., 109, 319-334.

Parkening, T.A. (1976). An ultrastructural study of implantation in the golden

hamster. I. Loss of the zona pellucida and initial attachment to the uterine epithelium. J.Anat., 121, 161-184.
Parr, E.L. (1973). Shedding of the zona pellucida by guinea pig blastocysts: An ultrastructural study. Biol. Reprod., 8, 531-544.
Reinius, S. (1967). Ultrastructure of blastocyst attachment in the mouse. Z. Zellforsch., 77, 257-266.
Samuel, C.A. (1971). The development of pig trophoblast in ectopic sites (Abstr.). J. Reprod. Fert., 27, 494-495.
Schlafke, S., and A.C. Enders (1975). Cellular basis of interaction between trophoblast and uterus at implantation. Biol. Reprod., 12, 41-65.
Sherman, M.I., and L.R. Wudl (1976). The implanting mouse blastocyst. In: G. Poste and G.L. Nicolson (Eds.), The Cell Surface in Animal Embryogenesis and Development. North-Holland Publishing Company, Amsterdam / New York / Oxford. pp. 81-125.
Smith, M.S.R., and S. Hartman (1974). Sarcoma cells as a blastocyst analogue in the mouse uterus (Abstr.). J. Reprod. Fert., 36, 465.
Spee, F. Graf von (1901). Die Implantation des Meerschweincheneies in die Uteruswand. Z. Morph. Anthropol., 3, 130-182.

Locomotory Behavior of a B16 Melanoma Variant Line Selected for Increased Invasiveness

I. R. Hart

Cancer Biology Program, NCI Frederick Cancer Research Center, P.O. Box B, Frederick, Maryland 21701, U.S.A.

ABSTRACT

To determine whether increased malignancy can be correlated with increased cell locomotion, we assayed the migration rates of two tumor lines with differing spontaneous metastatic behavior, using the agarose droplet technique. Enhanced malignancy was not associated with increased cellular mobility. However, when various normal cells were included in the assay, they exerted a selective effect upon the migration rates of the two tumor lines. The data indicate that normal host tissue might influence the pattern and the composition of secondary tumor growth.

KEYWORDS

Invasion; metastasis; tumor cell locomotion; agarose droplet; normal cell influence; heterogeneity.

INTRODUCTION

The invasive capacity of malignant neoplasms frequently is attributed to the increased translocative motility of individual tumor cells. Although little direct evidence is available to support such an hypothesis (Straüli and Weiss, 1977), the question can be approached in an indirect manner. Malignant tumor cells might be expected to exhibit greater locomotion *in vitro* than their benign counterparts, if increased motility were an important *invasive* mechanism.

Gershman and colleagues (1978), using SV-40 transformed hamster cells, were able to correlate increased tumorigenicity with increased cell mobility. Conversely, Varani and coworkers, using a murine tumor system and a different assay to measure cell migration, were unable to show an association between increased cell locomotion *in vitro* and more malignant behavior *in vivo* (Varani, Orr and Ward, 1978; 1979). In our study, we report the *in vitro* migratory capacity of two lines of the B16 melanoma selected for differing invasive behavior. These variant cell lines were assayed for locomotory behavior both alone and in the presence of a variety of normal cells in an attempt to understand the role of cell mobility in metastatic spread.

MATERIALS AND METHODS

Tumor Cell Lines

Two variants of the B16 melanoma, syngeneic to C57BL/6 mice, were used in this study. The B16-F10 cell line was selected for its lung colonizing ability following intravenous injections (Fidler, 1973), while the B16-BL6 was selected from this line for an increased invasive capacity by repeated passage through urinary bladders, of C57BL/6 mice, maintained in organ culture (Poste, Hart and Fidler, 1979; Hart, 1979).

Normal Cell Lines

Normal epidermal cells were isolated from C3H mouse tail skin by the method described by Steinmuller and Wunderlich (1976) and grown into cell lines designated TSII (tail skin epidermis). These nontumorigenic cells were a gift of Dr. M. Kripke, Frederick Cancer Research Center. Bovine endothelial cells were derived from pieces of calf aorta as described by Schwartz (1978). NIH 3T3 fibroblasts were a gift from Dr. J. N. Ihle, Frederick Cancer Research Center.

Cultures were maintained on plastic in Eagle's minimal essential medium supplemented with 10% fetal calf serum (FCS), sodium pyruvate, nonessential amino acids, L-glutamine and two-fold vitamin solution (CMEM) (Flow Laboratories, Rockville, MD). The cell lines were examined for and found free of mycoplasma and the following murine viruses: reovirus type 3, pneumonia virus of mice, K virus, Theiler's virus, Sendai virus, minute virus of mice, mouse adenovirus, mouse hepatitis virus, lymphocytic choriomeningitis virus, ectromelia virus and lactate dehydrogenase virus (Microbiological Associates, Walkersville, MD).

Migration Assay

Migratory activity of the two variant cell lines was measured using the assay described by Varani, Orr and Ward (1978). Briefly, subconfluent cultures of both cell lines were harvested and adjusted to form single cell suspensions consisting of 1×10^6 viable cells/ml CMEM (viability assessed by trypan blue exclusion). Equal aliquots of the suspensions were centrifuged, the supernates were discarded, and the cell pellets were resuspended in 0.2% agar in CMEM (0.3 ml agar/0.1 cell pellet). A drop of the agar/cell mixture (1-2 μl) was placed in the center of a Microtiter II well (Becton, Dickinson and Co., Oxnard, CA) and allowed to solidify at 4°C for 10 minutes when 0.1 ml chilled (4°C) CMEM was added to the well. The cultures were incubated in an humidified atmosphere, 5% CO_2 at 37°C, and the distance migrated by the tumor cells was measured at 24, 48 and 72 hours. An inverted microscope equipped with a linear micrometer (American Optical, Buffalo, NY) was used to measure the distance between the leading edge of the expanding corona of tumor cells and the edge of the agarose droplet.

In some experiments, as detailed in the Results section, the variant lines were either pretreated with cytochalasin B (CB) prior to assaying the migratory rate or were assayed in the presence of a cell pellet of one of the normal cell lines.

Treatment with CB was achieved by harvesting and adjusting the cells as described above and adding CB (5 mg dissolved in 1 ml dimethylsulfoxide) to give a final concentration of 10 μg CB/ml. The cell suspension was then incubated at 22°C for 30 minutes. Following this incubation period the agar/cell pellet was prepared and plated as described above.

In some experiments the migratory activity of the two tumor variant lines was assayed in the presence of an agar/cell pellet containing normal cells. These normal cells were harvested and plated as described for tumor cells and the two agar/cell droplets were placed in the same wall.

RESULTS

In Vivo Behavior of Tumor Variants

Representative results obtained by injecting equal numbers of B16-F10 and B16-BL6 cells into age- and sex-matched (C57BL/6 X C3H) F1 hybrid mice can be seen in Table 1. The B16-F10 cell line consistently produced a greater yield of lung nodules than the B16-BL6 cell line when equal numbers of cells were injected intravenously, but as seen in Table 2 the B16-BL6 variant was the more spontaneously metastatic cell line when injected either subcutaneously or intramuscularly.

TABLE 1 Lung Nodule Formation Following Intravenous Injection of B16 Variants

Tumor Cell Line	Numbers of Lung Nodules* Median (Range)
B16-F10	131 (39-200)
B16-BL6	64 (39-85)

*Mice were injected via the tail vein with 5 X 10^4 viable cells in 0.2 ml Hanks' balanced salt solution (HBSS) and autopsied 3 weeks later. There were 10 mice per group.

TABLE 2 Spontaneous Metastases from B16 Variant Cell Lines Injected Subcutaneously

Tumor Cell Line	Number of Mice with Pulmonary Metastases*
B16-F10	1/18 (6%)
B16-BL6	8/16 (50%)

*Mice injected subcutaneously in external ear with 5 X 10^4 viable cells in 0.1 ml HBSS. The ear was amputated 3 weeks later and mice were autopsied 3 weeks thereafter.

In Vitro Migratory Rates

A representative experiment comparing the in vitro migration of B16-F10 and B16-BL6 cells is shown in Fig. 1. Using this assay, B16-F10 cells are always slightly more motile than B16-BL6 cells over a 3-4 day period.

Treatment with CB reduced the observed motility of both the tumor cell lines (Fig. 2). On days 2, 3 and 4 the mobility of B16-F10 cells had been reduced by 43, 32 and 9%, respectively, whereas similar treatment reduced the motility of B16-BL6 cells by 35, 33 and 14% at the same time periods.

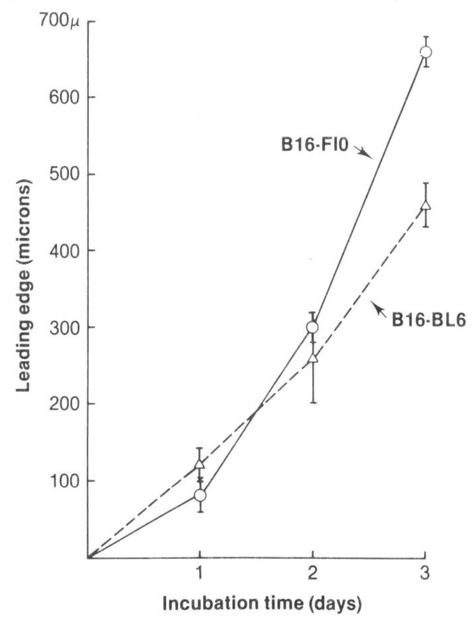

Fig. 1. The migratory rates of B16 variant lines. Each value represents the mean of 16 individual measurements. Vertical bars represent ± S.D.

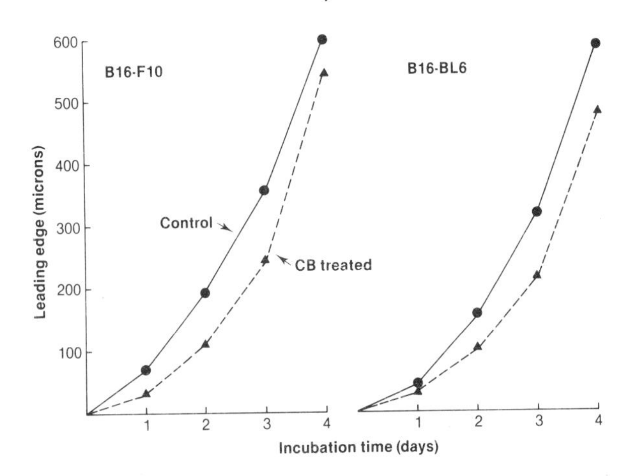

Fig. 2. Inhibition of tumor cell motility by cytochalasin B. Points represent the mean of 16 individual measurements. S.D. $\leq$ 5%.

The migratory rates of B16-F10 and B16-BL6 cells in the presence of bovine endothelial, TSII epidermal and 3T3 fibroblasts cells are shown in Fig. 3.

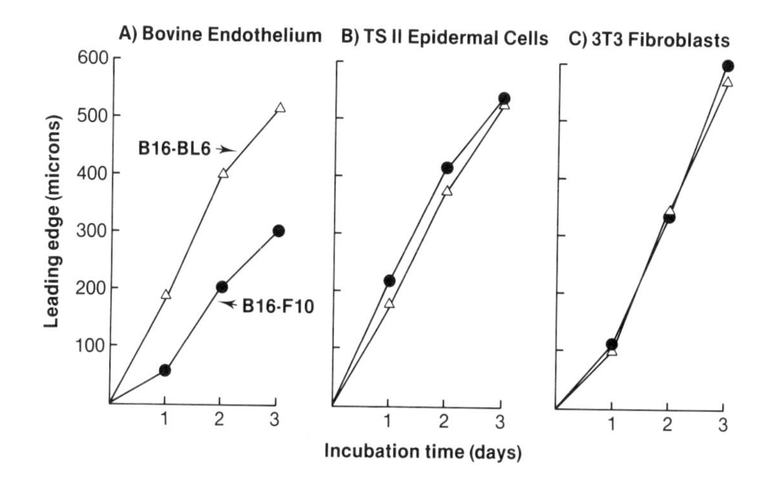

Fig. 3. Effect of nontumorigenic cells on tumor cell motility. Migration rates of the two variants were assayed in the presence of the designated normal cells as described in the text. Each point is the mean of 16 individual measurements. S.D. $\leq 5\%$.

The migration of both B16-F10 and B16-BL6 cells was not affected by the presence of either TSII epidermal cells or 3T3 fibroblasts in adjacent agarose droplets. However, the inclusion of an agarose droplet containing bovine endothelial cells in the same well as the migrating tumor cells revealed marked differences between the two variant lines. The presence of endothelial cells significantly depressed ($p \leq 0.001$) the motility of the B16-F10 but not that of the B16-BL6 cells.

DISCUSSION

The process of tumor invasion involves the movement of neoplastic cells, with or without attendant tissue destruction, into areas previously occupied by normal host cells. One of the many mechanisms (for review see Fidler, Gersten and Hart, 1978) thought to be responsible for this characteristic of malignant tumors is the motility of individual cells shed from the primary tumor focus (Strauli and Weiss, 1977). Few comparisons of migration rates of tumor cells and homologous normal cells have been made (Easty and Easty, 1976) although such an approach is essential to determine the role played by cell locomotion in malignant invasion. Indeed, the need to compare tumor cells with normal (or seemingly normal) cells could be obviated by comparing tumor cell lines of differing metastatic behavior. This approach has been used with hamster and murine tumor systems, but no clear association between malignancy and locomotory capacity has been demonstrated (Gershman, Katzin and Cook; 1978; Varani, Orr and Ward, 1979).

No direct correlation between metastasis and cell mobility can be demonstrated in the B16 melanoma using the assay of Varani and co-workers (Varani, Orr and

Ward, 1978). B16-F10 was consistently more mobile than B16-BL6, which was the more spontaneously metastatic cell line. We have previously shown that the microfilament-disrupting drug CB can decrease cell motility and markedly alter metastatic patterns of the B16 melanoma (Hart, Raz and Fidler, 1979). Cytoskeletal elements are probably of fundamental importance in mediating some aspects of the invasive and metastatic processes. However, preincubation of both B16-F10 and B16-BL6 cells with equal amounts of CB brought about an equal reduction in the motility of these two lines *in vitro*. The microfilament composition of the two variants, as assessed by this functional test, appears to be similar. It is unlikely that cell motility *per se* accounts for the observed differences in spontaneous metastasis in this tumor system.

Motility rates of the two tumor variant lines were assayed in the presence of different normal cells using the agarose explant technique. The presence of endothelial cells depressed the motility of the B16-F10 but not the B16-BL6 cells. The heterogeneity of the B16 melanoma with regard to metastatic potential has been clearly demonstrated (Fidler 1973;1978; Fidler and Kripke, 1977). The data reported here suggest that such heterogeneity may extend to the response of tumor cells to normal cells. Cells of diverse origin may exert a selective effect on the mobility of subpopulations of tumor cells. Should such an *in vitro* effect be exerted *in vivo* it is conceivable that the tissues of the tumor-bearing host could to some extent determine patterns of tumor dissemination and the nature of the tumor cell population that finally grows in distant organs. Further work is in progress to determine whether such mechanisms do play a role in the dissemination of the B16 melanoma.

ACKNOWLEDGEMENT

Research supported by the National Cancer Institute under Contract No. N01-CO-75380 with Litton Bionetics, Inc.

REFERENCES

Easty, G. C. and D. M. Easty (1976). Mechanisms of tumor invasion. In T. Symington and R. L. Carter (Eds.) *Scientific Foundations on Oncology* Heinemann, London, pp. 167-172.
Fidler, I. J. (1973). Selection of successive tumor lines for metastasis. *Nature (Lond.) New Biol. 242*, 148-149.
Fidler, I. J. (1978). Tumor heterogeneity and the biology of cancer invasion and metastasis. *Cancer Res. 38*, 2561-2660.
Fidler, I. J. and M. L. Kripke (1977). Metastasis results from pre-existing variant cells within a malignant tumor. *Science 197*, 893-895.
Fidler, I. J., D. M. Gersten and I. R. Hart (1978). The biology of cancer invasion and metastasis. *Adv. Cancer Res. 28*, 149-250.
Gershman, H., W. Katzin and R. T. Cook (1978). Mobility of cells from solid tumors. *Int. J. Cancer 21*, 309-316.
Hart, I. R. (1979). Selection and characterization of an invasive variant of the B16 melanoma. *Am. J. Path.* (in press).
Hart, I. R., A. Raz and I. J. Fidler. (1979). The role of cytoskeletal elements in tumor dissemination. *J. Natl. Cancer Inst.* (submitted for publication).
Poste, G. H., I. R. Hart and I. J. Fidler. (1979). Manuscript in preparation.
Schwartz, S. M. (1978). Selection and characterization of bovine aortic endothelial cells. In Vitro 14, 966-980.
Steinmuller, D. and J. R. Wunderlich. (1976). The use of freshly explanted

mouse epidermal cells for the *in vitro* induction and detection of cell mediated cytotoxicity. *Cell. Immunol. 24*, 146-153.

Strauli P. and L. Weiss. (1977). Cell locomotion and tumor penetration. *Europ. J. Cancer 13*, 1-12.

Varani, J., W. Orr and P. A. Ward (1978). A comparison of the migration patterns of normal and malignant cells in two assay systems. *Am. J. Path. 90*, 159-171.

Varani, J., W. Orr and P. A. Ward. (1977). Comparison of subpopulations of subpopulations of tumor cells with altered migratory activity, attachment characteristics, enzyme levels and *in vivo* behavior. *Europ. J. Cancer 15*, 585-592.

Some Further Characteristics of Human Mesothelioma Cells

E. Nissen, W. Arnold, H. Weiss and St. Tanneberger

Academy of Sciences of the GDR, Central Institute of Cancer Research, 1115 Berlin, GDR

ABSTRACT

Ascitic tumour cells from a female patient with peritoneal carcinomatosis have been cultivated in vitro. The determination of contact inhibition of movement by means of overlapping ratios and the lactic dehydrogenase (LDH) isoenzyme patterns prior to and after transplantation into nude mice are described. DNA distribution patterns and drug sensitivity were compared after several passages in vivo (AP 14,AP 18;S 3) and in vitro (K 8,K 25,K 28,K 39,K 43; SC,SC 3,SC 14,SC 18).

Cells growing in nude mice show a somewhat higher rate of observed-to-expected overlaps and different LDH isoenzyme patterns. DNA distribution patterns under in vitro conditions show no changes. The G 1/0-peak of the ascitic tumour cells (AP 14,AP 18) revealed a shift to the left. The drug sensitivity was constant during in vitro culture with the exception of 5-Fluorouracil. Differences in sensitivity to Methotrexate and 5-Fluorouracil were observed after transplantation into nude mice. The possible reasons of instability of DNA distribution patterns and drug sensitivity are discussed.

KEYWORDS

Cell culture; DNA distribution pattern, drug sensitivity,LDH isoenzyme pattern, contact inhibition.

INTRODUCTION

Studies on the problem of the stability of biological tumour properties are important both for basic and clinical cancer research. In the following a set of experiments is discussed concerning the question in how far cellular properties are changeable by the biological environment and time. Some characteristics (DNA distribution patterns, chromosome analysis, growth curves) are published by Nissen and colleagues (1979); here we will describe the determination of contact inhibition of movement by means of overlapping ratios, LDH isoenzyme patterns prior to and after transplantation into nude mice and the comparison of DNA distribution patterns and drug sensitivity after several passages in vivo and in vitro. Figure 1 shows the principle of the study. Ascitic tumour cells from a female patient with peritoneal carcinomatosis have been cultivated in vitro. DNA distribution patterns and drug sensitivity were determined of the 8th passage in vitro (K 8). Cells of the 25th in vitro passage (K 25) were also characterized and i.p. transplanted in nude mice, where they grew in ascitic form and as solid tumours. The 3rd solid passage (S 3) has been cultivated in vitro again (SC). In the study cells of in vitro cultures (K 8, K 25,K 28,K 39,K 43; SC,SC 3,SC 14,SC 18) and of two nude mice passages, the 14th (AP 14) and the 18th (AP 18) have been compared with respect to the DNA distribution pattern and drug sensitivity in vitro.

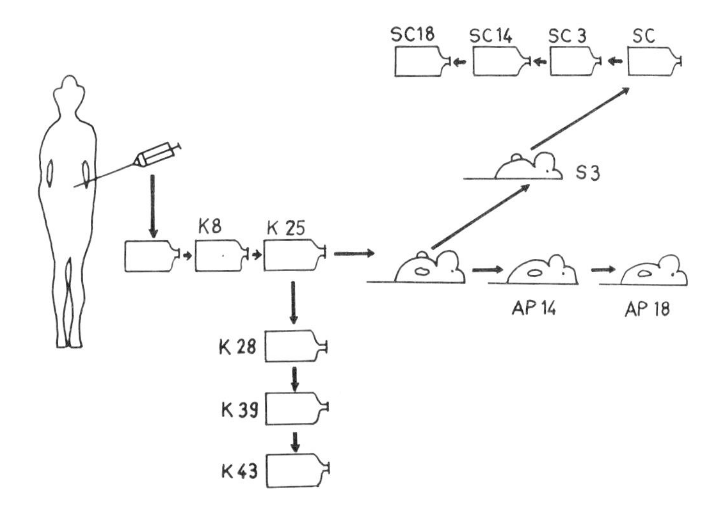

Fig. 1 Principles of the study on human tumour cell explantation into an in vitro system (cell culture) and an in vivo system (nude mice xenografting).

METHODS

Details of chemicals, determination of DNA distribution patterns and drug sensitivity testing are published by Nissen and colleagues (1979). For the quantitative evaluation of the contact inhibition of movement the method described by Projan and co-workers (1973) was used. LDH isoenzyme patterns were measured by agar-gel electrophoresis (Pecse,1977).

RESULTS and DISCUSSION

The results of cell movement determined by calculation of overlaps of cultivated cells are summarized in Table 1. Cells growing in nude mice show a somewhat higher rate of observed-to-expected overlaps. Differences are not significant. In vitro cultures of

TABLE 1 Data for Calculation of Overlaps of Cultivated Cells prior to and after Transplantation in Nude Mice

Cell type	Effect. nucl.	E	O	O/E (%)
K 25	0.30	6.1	0.5	7.4
AP 14	0.19	2.6	0.4	14.1
AP 18	0.33	5.3	0.5	10.3
SC	0.25	2.8	0.5	16.2

Effect. nucl. effective nuclear area
E expected number of overlaps
O observed number of overlaps

cells grown in nude mice contain a higher percentage of fibroblast-like cells. Maybe this type of cells shows a higher rate of overlaps than epithelially growing cells. In comparison to early studies (Projan,1973) the tumour cell populations described here do not differ from normal cell cultures.

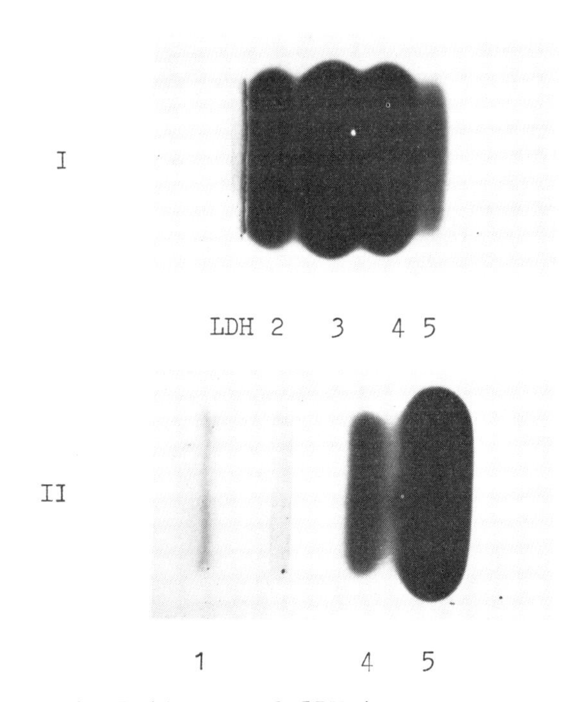

Fig. 2 Patterns of LDH isoenzymes prepared from K 25 cell extract (I) and AP 14 cell extract (II).

The K 25-cells showed 3 strongly (LDH 2,3,4) and 1 weakly staining (LDH 5) band. In contrast to this pattern AP 18-cells showed 1 strong (LDH 5) and 1 weak (LDH 4) band. The bands run more quickly. A possible explanation could be the differentiation process of cells. LDH isoenzyme patterns like AP 14-cells is an example for highly differentiated cells. Ziegenbein and colleagues (1974) has obtained similar results.

DNA distribution patterns of several both in vitro passages of K-cells, SC-cells (Fig. 3) and in vivo passages (Fig. 4) were summarized.

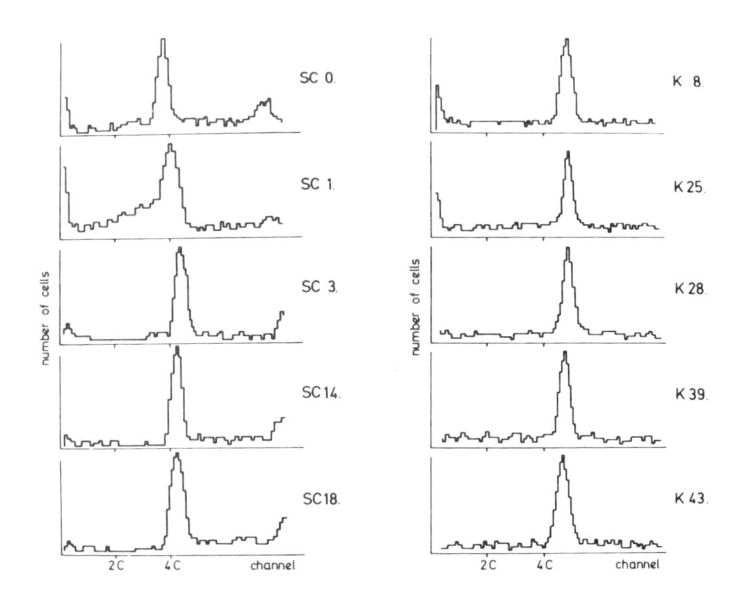

Fig. 3 DNA distribution patterns of human tumour cells growing under in vitro conditions.

In vitro conditions did not change DNA distribution patterns. The standardized in vitro system and constant environmental conditions (medium, serum, pH etc.) could be responsible.

In contrast to these findings, changes are observable on transplantation into nude mice. As shown in Fig. 4 the G 1/0-peak of the ascitic tumour cells (AP 14, AP 18) reveal a shift to the left compared with the in vitro cultivated cells K 25 and SC.

Changes in the ploidy cannot be explained by a stronger immunogenicity of these cells because successful xenogenic transplantation was performed into athymic nude mice. An explanation could be selection. Maybe cells are to a different degree capable of proliferating and/or invading the host environment. Nutritional and other humoral factors, e.g. hormones, might play a role in this selection in vivo. These findings correlate with experiments of Laerum and co-worker (1976).

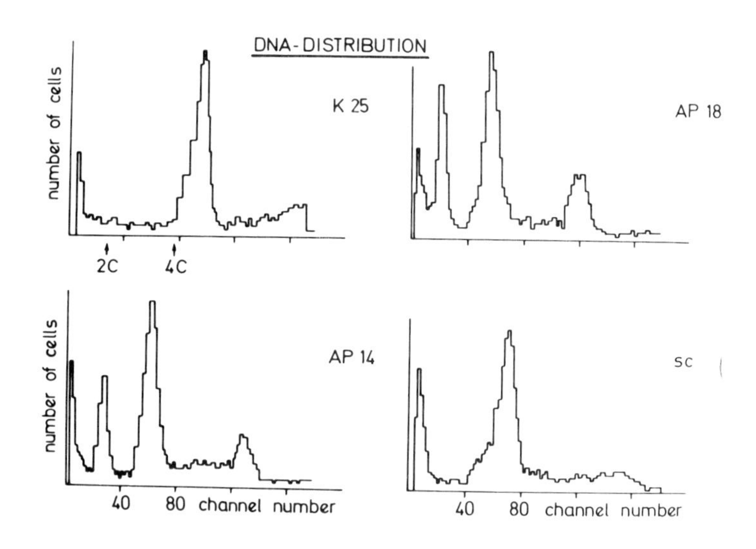

Fig. 4 Pulse cytophotometry measured DNA distribution of human tumour cells growing under in vitro conditions (K 25,SC) and in vivo (AP 14,AP 18).

The results of drug sensitivity testing of several in vitro passages of K-cells are summarized in Table 2.

TABLE 2 Drug Sensitivity of Ascitic Cells of the Patient (K-cells)

Passages /ug/ml	4-HPC 10.0	D/R 1.0	F 20.0	V 0.2	M 1.8
K 8	-	99.6 +)	0	82.6	0
K 25	97.3	96.3	43.9	97.5	0
K 28	95.8	96.6	33.9	98.8	0
K 39	97.9	96.5	29.6	98.8	0
K 43	96.9	97.1	36.9	98.9	0

+) % inhibition of 3H-thymidine incorporation

No differences in drug sensitivity have been found against 4-Hydroperoxycyclophosphamide (4-HPC), Daunoblastine (D), Rubomycine (R) and Vinblastine (V). Methotrexate (M) was ineffective. Changes in drug sensitivity are visible against 5-Fluorouracil (F) more in SC-cells (Table 3) than in K-cells. Early passages of solid material (SC,SC 3) show the same sensitivity pattern as the first determination of K-cells (K 8,Table 2). This could be referred to the fact

TABLE 3 Drug Sensitivity of Solid Tumour Material and Ascitic Cells after Transplantation into Nude Mice

Tumour material	Passages	4-HPC	D/R	F	V	M
solid						
organ cultures	S 3	95.3+)	35.1	0	67.7	0
cell cultures	SC	93.7	83.6	0	89.2	0
	SC 3	95.4	98.2	14.8	92.0	0
	SC 14	96.5	91.5	69.1	98.2	0
	SC 18	95.7	98.5	86.5	99.7	0
ascitic cells	AP 14	94.7	82.8	95.8	96.4	0
	AP 18	99.5	81.9	96.7	78.9	68.1

+) % inhibition of 3H-thymidine incorporation

that this material is not yet so well adapted to the in vivo conditions. Thus the cells of the ascitic fluid were passaged 14 and 18 times but solid tumours only three times. In contrast to the findings of in vitro cultures, the drug sensitivity of AP 18-cells was changed. The resistance of Methotrexate disappeared and the sensitivity to Vinblastine was slightly decreased.

Selection processes are well known to occur during establishment of cell cultures. Only those cells are growing in tissue culture which are able to adaptation. Looking for a correlation between biological behavoir and drug sensitivity we found that changes in sensitivity to antineoplastic agents could be explained possibly by a selection of different subpopulations of the human tumour material. A human neoplastic cell population is mostly a mixture of several tumour cell clones. Every cell clone has specific characteristics (Pályi,1977) and in the course of tumour development sublines are changeable by means of mutation, differentiation, selection and/or other processes.

In summary the following conclusions can be drawn from these and earlier studies:

The individual antineoplastic drug sensitivity of human tumours on cellular level is related to the ploidy and the growth rate of the population. The biological properties including the drug sensitivity of human tumour cell populations can be changed by the biological environment of the cells and time. The changes in the properties can be the consequence of cellular changes or of changes in the composition of the normally multiclonal, mixed cell population by selection. On the basis of our data no answer can be given to this question. Differences in the biological properties of the different cell clones in a mixed population including their drug sensitivity are well proved (Pályi,1977). The findings described here can contribute to the explanation of such phenomena as secondary drug resistance of human tumours or the different drug response of primary tumours and metastases often observed. Furthermore the lack of useful drug prediction tests obviously at this time in spite of intensive international efforts (Tanneberger,1978) can be explained.

ACKNOWLEDGEMENT

We thank Dr. R. Ziegenbein und H. Goebel for determination of LDH isoenzymes.

REFERENCES

Laerum, O.D., and S.J. Mork (1976). DNA distributions in aneuploid tumor cell lines under different culture and transplantation conditions. Pulse-Cytophotometry,1, 250-260.
Nissen, E., W. Arnold, H. Weißss, and St. Tanneberger (1979). Biological characterization of a human mesothelioma in nude mice II. Some characteristics of cells cultivated in vitro prior to and after transplantation into nude mice. Arch.Geschwulstforsch.,49, in press.
Pályi, I., E. Olah, and J. Sugar (1977). Drug sensitivity studies on clonal cell lines isolated from heteroploid tumour cell populations I. Dose response of clones growing in monolayer cultures. Int.J. Cancer, 19, 859-865.
Pesce,A.J., H.C. Bubel, L. Di Persio, and J.G. Michael (1977). Human lactic dehydrogenase as a marker for human tumor cells grown in athymic mice. Cancer Res.,37,1989-2003.
Projan, A., and St. Tanneberger (1973). Some findings of movement and contact inhibition of human normal and tumour cells in vitro. Europ. J. Cancer,9, 703-708.
Tanneberger,St., E. Nissen, and W. Schälike (1978). Prediction of drug efficacy potentialities and limitations. XI International Cancer Conference, Buenos Aires.
Ziegenbein,R., C.N. Schremmer, and Ch. Fengler (1974). LDH-Isoenzyme in Mamma-Karzinomgeweben unterschiedlicher Reifegrade. Z. Exper. Chirurgie,7, 228-234.

Invasiveness of Malignant Mouse Fibroblasts *in vivo*

C. Meyvisch and R. Van Cauwenberge

Department of Experimental Cancerology, Clinic for Radiotherapy and Nuclear Medicine, Academic Hospital, B-9000 Ghent, Belgium

ABSTRACT

In this comparative in vitro-in vivo study we test the hypothesis that the invasive capacity of tumour cells is necessary for the take of a transplant. Spheroidal aggregates of MO_4 cells, shown to be invasive in vitro, were treated before implantation with several drugs, that all block the growth of the tumour cells but that have a different effect on their migration. Subcutaneous inoculation of aggregates treated with 5-fluorouracil produces the same number of tumours as the controls. With mitomycin the tumour take is greatly reduced as it is the case after treatment of the aggregates with the microtubule inhibitors vincristin or nocodazole. The morphology of treated MO_4 cells, migrating from aggregates explanted on glass, suggests that the length of the period before appearance of a palpable tumour nodule is related to the capacity of the cells to perform directional migration.

KEYWORDS

Tumorigenicity; latency period; MO_4 aggregates; directional migration; growth; 5-fluorouracil; mitomycin; nocodazole; vincristin.

INTRODUCTION

We have developed an animal model (Meyvisch and Mareel, 1979) that allows comparison of invasion in vivo and in vitro. The course of early invasion from spheroidal aggregates of MO_4 cells (virally transformed malignant C_3H mouse fibroblasts) after subcutaneous inoculation has been followed on serial histological sections. Solitary invasive MO_4 cells penetrating into the surrounding host tissues are observed during the first three days. In this period MO_4 cells that remain in the aggregate are destroyed by an inflammatory host reaction. These observations have led us to conclude that only a small fraction of invasive MO_4 cells will grow into a palpable tumour at a site where a favourable microenvironment is present. This implies that invasion of implanted malignant cells is important for the take of the tumour. According to Mareel and De Brabander (1978) the capacity of directional migration is necessary for the invasiveness of malignant cells. The growth of individual MO_4 aggregates in shaker culture can be followed (Mareel, Kint and Meyvisch, 1979) and an assay for directional migration of these cells has been described by Storme and Mareel in this volume. From their experiments it appears that various agents that affect the growth of the aggregates of MO_4 cells have different effects on the directional migration and invasive capacity of these cells in vitro. Preliminary experiments using both methods have shown that 1 µg/ml 5-fluorouracil or 0.1 µg/ml mitomycin stops growth of the aggregates within 48 hours but allows directional migration. Nocodazole or vincristin abolishes directional migration by interference with the cytoplasmic microtubular complex and inhibits malignant invasion in vitro (Mareel and De Brabander, 1978). At a concentration of 1 µg/ml during 48 hours both drugs are able to block the growth of the MO_4 aggregates in vitro. Aggregates of MO_4 cells pretreated in shaker culture during 48 hours with one of these four drugs provide us with non-growing tumour cell populations with different capacities of directional migration.

GROWTH OF PRETREATED AGGREGATES IN VITRO

If the growth of pretreated aggregates of MO_4 cells in vitro has to be correlated with the formation of tumours in vivo, the resumption of growth in vitro after exposure to the drugs has first to be checked.
Individual MO_4 aggregates are treated in shaker culture using Eagle's minimum essential medium (MEM, Flow Laboratories, Ltd, Irvine, Scotland) plus 10% fetal bovine serum and 0.05% glutamine, supplemented with 5-fluorouracil (Roche Laboratories, Nutley, N.J.) : 1 µg/ml; or with mitomycin C (Christiaens, Brussels, Belgium) : 0.1 µg/ml; or with vincristin (Lilly, Brussels, Belgium) : 1 µg/ml; or with nocodazole (Janssen Pharmaceutica, Beerse, Belgium) : 1 µg/ml. Nocodazole is dissolved in DMSO at a concentration of 5 mg/ml. After 48 hours the drugs are washed out and the medium is refreshed every 24 hours. Each aggregate is photographed daily during 12 days and the diameter of the aggregates is plotted as percentage of the original diameter. Fig. 1 shows that only aggregates treated with mitomycin are blocked irreversably, whilst aggregates treated with 5-fluorouracil, nocodazole or vincristin resume growth at approximately the same rate after a short lag period.

DIRECTIONAL MIGRATION FROM TREATED AGGREGATES IN VITRO

After two days treatment of the MO_4 aggregates in shaker culture the drug is washed out and single aggregates with the same diameter (0.3 mm) are explanted on coverslips in Leighton tubes. The medium is refreshed every day. The mean diameter of the circular area covered by the cells that migrate from the aggregate in absence of the drugs is measured during seven days (Fig. 2).
Using this method the migration of MO_4 cells from all treated aggregates seems very similar. However, the morphology and the number of cells that are present outside the aggregate differ greatly with various treatments. In controls and in aggregates treated with 5-fluorouracil or with mitomycin, cells that have migrated from the aggregate are spindle shaped and radially oriented (Fig. 3). Contrary after treatment with nocodazole or vincristin the cells laying on the glass next to the aggregate are polygonal, usually containing several nuclei and radial orientation cannot be recognized. The number of cells that have migrated from the aggregates after 3 days is about 70% (5-fluorouracil), 35% (nocodazole and vincristin) and 10% (mitomycin) of the controls. These data indicate that after treatment with

5-fluorouracil or mitomycin, MO_4 cells conserve their capacity of directional migration, which is not the case after treatment with the microtubule inhibitors.

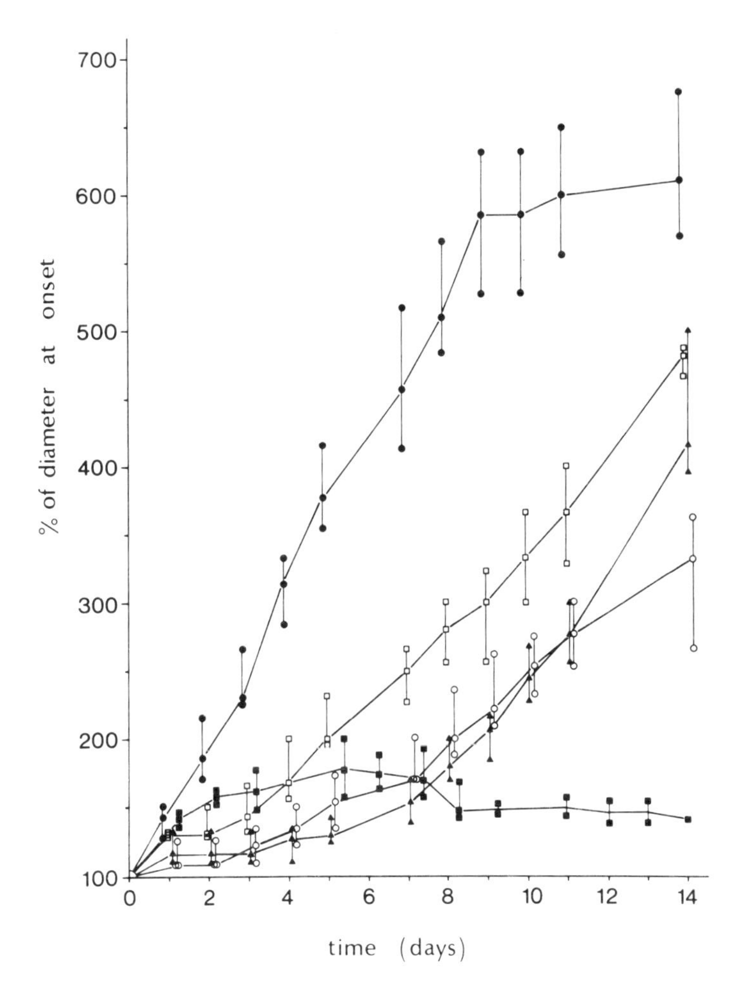

Fig. 1. Growth in vitro of aggregates treated during 2 days with 5-fluorouracil (▲-▲), mitomycin C (■-■), nocodazole (□-□) or vincristin (o-o), washed and cultured further in medium without drug. (●-●) : control.

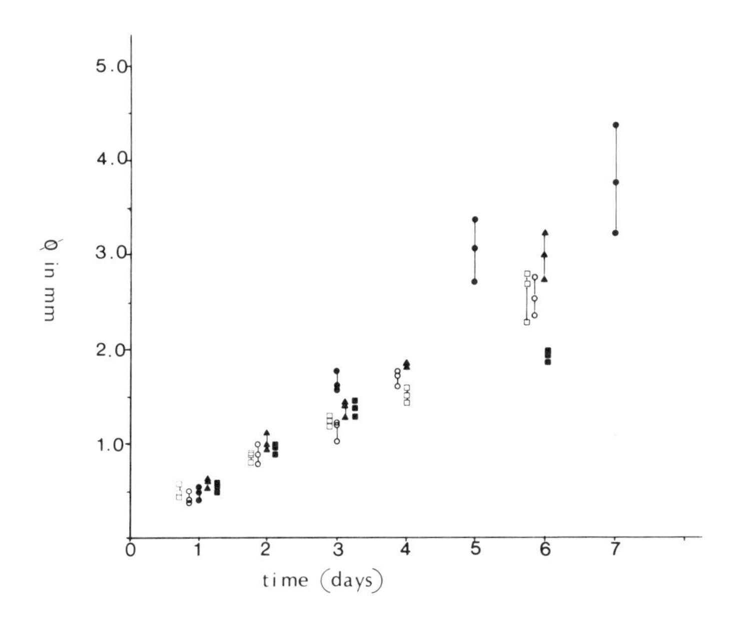

Fig. 2. Migration of MO_4 cells from aggregates treated during 2 days with 5-fluorouracil (▲-▲), mitomycin (■-■), nocodazole (□-□) or vincristin (○-○) and explanted on glass in medium without drugs. Ordinate : diameter of the circular area covered by the cells that leave the aggregate. (●-●) : control.

TUMORIGENICITY IN VIVO

A single MO_4 aggregate, pretreated for 48 hours with one of the four drugs is implanted subcutaneously on the dorsal side of the auricle of the external ear of a female inbred C_3H/He mouse with the aid of a micropipette (MICROPETTOR, SMI Scientific Manufacturing Industries, Emeryville, Cal.). Previously a tunnel is made with a sharp stainless steel needle (diameter = 0,5 mm) from an incision at the base of the auricle towards its tip (Fig. 4).
For each drug ten animals are inoculated and the number of tumour takes and the latency period before appearance of a palpable nodule ($\pm$ 1 mm) is compared with a series of control animals inoculated with untreated MO_4 aggregates of the same size. Figure 5 shows that the aggregates treated with 5-fluorouracil from nearly the same number of tumours as the untreated controls whilst vincristin and mitomycin lower the tumorigenicity considerably. After nocodazole treatment no tumours are formed. The latency period before appearance of a nodule is the same after inoculation of 5-fluorouracil or mitomycin treated cells, as for the controls but for the microtubular inhibitor vincristin it is prolonged up to 30 days. In a preliminary series the MO_4 aggregates have been pretreated with nocodazole during only 24 hours. One of five animals inoculated has developed a tumour after 29 days.

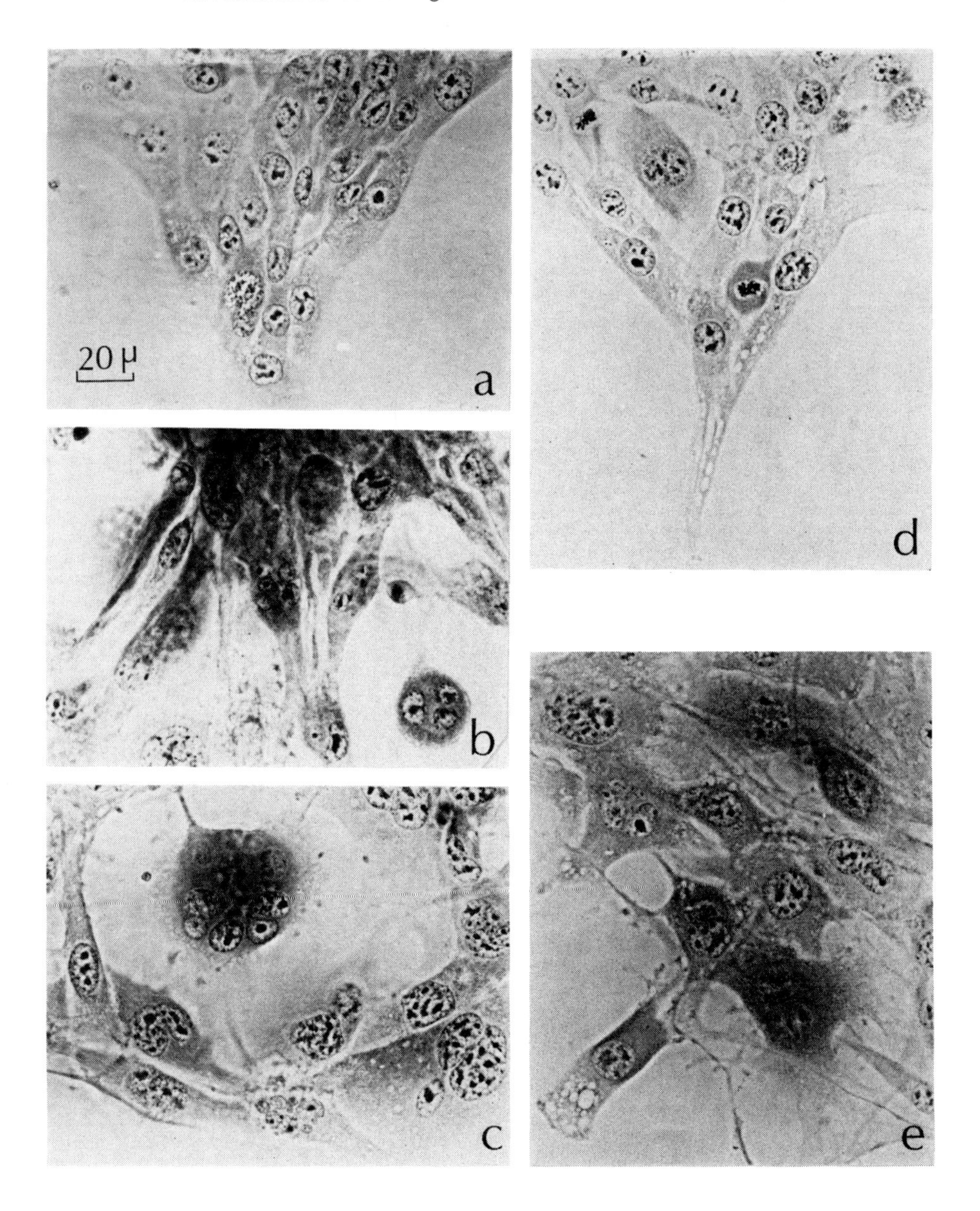

Fig. 3. Phase contrast photomicrographs of the outer rim of the circular area covered by MO_4 cells three days after explantation of untreated aggregates (a) and aggregates treated with mitomycin (b), nocodazole (c), 5-fluorouracil (d) or vincristin (e).

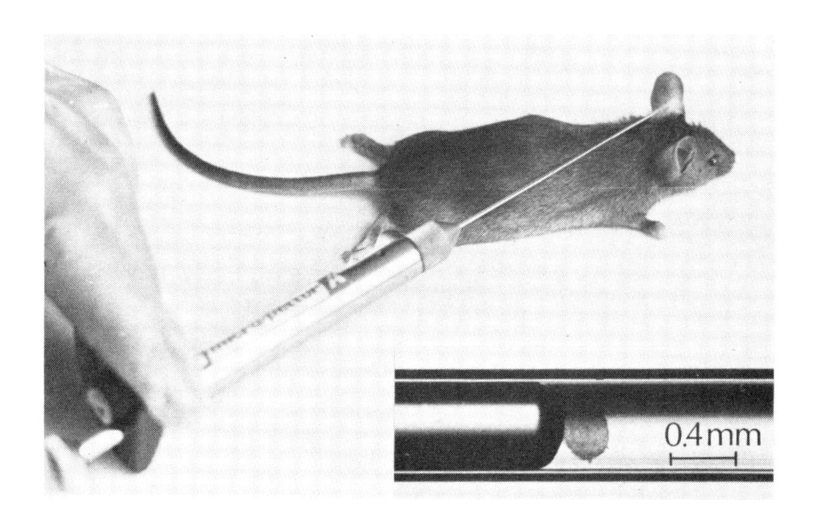

Fig. 4. Inoculation procedure. The inset shows the tip of the micropipette with plunger and MO_4 aggregate.

CONCLUSIONS

Comparison between tumorigenicity, growth and directional migration of MO_4 cells after treatment with various drugs supports our hypothesis that invasion of cells from an aggregate implanted subcutaneously is important for the development of a tumour.

Histology of aggregates of MO_4 cells fixed from 1 to 6 days after implantation into syngenic mice suggests that the tumour is formed from cells that invade early into the surrounding tissues, whereas MO_4 cells remaining at the site of inoculation are destroyed by the reaction of the host (Meyvisch and Mareel, 1979).

Storme and Mareel (this volume) have found a good correlation between the invasiveness of MO_4 cells and their capacity to perform directional migration on glass.

Hence, it can be expected that drugs which interfere with directional migration will inhibit tumorigenicity more than drugs which allow directional migration, although the effect of both on growth is similar. This is what is found in the present experiments. Tumorigenicity is high with a short latency period for aggregates treated with 5-fluorouracil, the cells of which resume growth and conserve their capacity of directional migration in vitro. Tumorigenicity is low with a long latency period for aggregates treated with microtubule inhibitors, the cells of which resume growth but have lost their capacity of directional migration at least for three days, a period that is critical for invasion in vivo (Meyvisch and Mareel, 1979).

The short latency period of two tumours formed by aggregates treated with mitomycin fits with these data. Formation of these tumours, however, indicates that the reversibility of the action of the drug on growth is different in vivo and in vitro. This might be ascribed to differences in the microenvironment. Another possibility is, that in certain conditions measurement of the diameter of an aggregate in shaker culture is not sufficient for the study of the proliferative capacity of a cell population (Yuhas and Li, 1978).

Although the present data support our hypothesis that invasion is important for

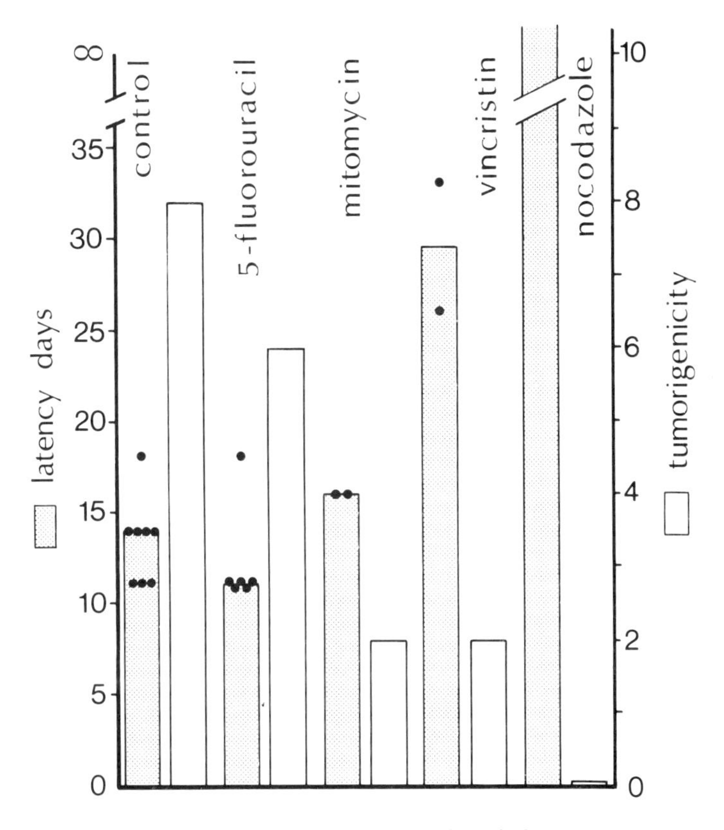

Fig. 5. In vivo tumorigenicity

formation of a tumour in vivo, a few alternative explanations remain to be examined. Amongst them alteration of the host reaction towards treated MO_4 cells as compared to controls deserves consideration. In this respect the histology of tumour formation from treated aggregates will be useful.

REFERENCES

Mareel, M.K., and M.J. De Brabander (1978). Effect of microtubule inhibitors on malignant invasion in vitro. J. Natl. Cancer Inst., 61, 787-792.

Mareel, M., J. Kint, and C. Meyvisch (1979). Methods of study of the invasion of malignant C3H-mouse fibroblasts into embryonic chick. Virchows Arch. B Cell Path., 30, 95-111.

Meyvisch, C., and M. Mareel (1979). Invasion of malignant C3H-mouse fibroblasts from aggregates transplanted into the auricles of syngenic mice. Virchows Arch. B Cell Path., 30, 113-122.

Yuhas, J.M., and A.P. Li (1978). Growth fraction as the major determinant of multicellular tumor spheroid growth rates. Cancer Res., 38, 1528-1532.

ACKNOWLEDGMENT

This work is supported by a Grant from the Kankerfonds van de Algemene Spaar- en Lijfrentekas and from the NFGWO, Brussels, Belgium.

The Role of Cell Movement in Tumor Invasion: a General Appraisal

P. Sträuli

Division of Cancer Research, Institute of Pathology, University of Zürich, Birchstrasse 95, 8050 Zürich, Switzerland

ABSTRACT

Movement of tumor cells in its stationary and translocative form can contribute to invasion, as demonstrated indirectly by histology and directly by microcinematography. In most instances, tumor cell movement as a factor of invasion operates conjointly with proliferation and lytic action. In its translocative form (locomotion), tumor cell movement is accomplished by three closely correlated mechanisms: generation of driving force, adaptation of cell shape, and formation of attachment sites.

KEYWORDS

Tumor invasion; stationary cell motility; translocative cell motility; cell locomotion; driving force generation; cell shape; cell adhesion; proliferation and invasion; enzyme action and invasion.

The advance of cancer elements into normal host tissues is the result of a complex interaction between tumor and host factors. To become experimentally exploitable, this interaction must be subdivided into functional units. On the side of the tumor, the main functions possibly contributing to invasion are proliferation, cell motility (movement), and lytic action. It is unlikely that each factor alone is able to achieve invasion: in most instances, the three functions can be expected to operate conjointly, although one of them may prevail. Expression of tumor activities is modified (hampered or promoted) by host reactions which rather appear as unconnected homeostatic mechanisms than as an integrated defense against invasion.

If we now consider cell movement, we have first to furnish proof of its contribution to invasion (Sträuli & Weiss, 1977).

Evidence of an indirect nature is provided by histology, but requires complete sets of serial sections or even three-dimensional reconstruction. Under these conditions, cancer cells are often found at localizations that can only have been attained by self-propulsion and not by passive transport. Attempts to demonstrate cell movement by static morphology are most promising where the whole scene of conjectured cell activities can be surveyed by histology. Such an approach has successfully been applied to the study of cell migrations in the embryo (Le Douarin, 1976), but can much less easily be used for proving the occurrence of cell movement in the large invasion zones of primary animal and human tumors. Where feasible, in vitro-models may prove helpful, as it is one of their advantages that due to their small dimensions they can be subjected to complete microscopic examination. Histological evidence for cell movement is supported by observations of cells in polarized configuration, as, in all likelihood, such cells were caught by fixation while locomoting (Felix, Haemmerli and Sträuli, 1978).

Direct evidence for the contribution of cell movement to invasion consists in microcinematographic recording of the pertinent cell activities. The classical approach utilizes transparent chambers inserted into and contiguous with living tissue. However, the difficulties of bringing together a chamber with good optical properties and an invading tumor are considerable, and few reliable data have been obtained so far. In the rabbit ear chamber, Wood (1958) was able to record the penetration through vessel walls and the extravascular locomotion of V 2 carcinoma cells introduced into ear arteries. Considering the intricacies of in vivo-models, we have now introduced a compromise technique: the mesentery infiltrated by cancer cells in vivo and subsequently subjected to microcinematography in vitro (Haemmerli and Sträuli, 1978). This new model is able to provide unequivocal evidence for cancer cell movement within a living tissue. Furthermore, it offers an excellent opportunity for studying the interaction of cell motility with other mechanisms of invasion.

Cell motility in general can be subdivided into a stationary and a translocative form (Haemmerli and Sträuli, in press). Histology and microcinematography predominantly demonstrate the translocative form, cell locomotion. But stationary cell movement also contributes to invasion.

A special type of stationary cell movement to be considered here is cytokinesis. After cell cleavage, two (or more) cancer cells have to adapt to the site previously occupied by one cell. This requires positional shifting of the daughter cells which, at the beginning, is stationary, but on principle can turn into locomotion. In differentiated solid cancer with maintained tissue cohesion, the transition from cytokinesis to locomotion as a cell release mechanism appears to be rare, and invasion is predominantly proliferative ("invasive growth"): new tumor substance is continuously added to the penetration front (and often also to more central parts of the tumor where it generates outward pressure). In cancer with decreased or abolished tissue cohesion, a functional coupling of cytokinesis and locomotion is possible, but not imperative. Cells will probably start locomoting in other phases of the cycle as well (with the exception of mitosis itself), and proliferation is not the leading mechanism of invasion.

Another type of stationary motility consists in the projection of cytoplasmic extensions. They can be inserted between host cells; this allows cancer cells to exert a lever action and results in minute disruptive effects. The evidence for this process is largely indirect (static-histological), but there are a few relevant microcinematographic observations

(Ambrose and Easty, 1973). In electron micrographs, the basal lamina adjacent to cancer cell extensions is occasionally found to be displaced or pierced (Frithjof,1972). The latter observation raises the question whether the thrust of stationary cell motility alone is able to produce gaps in the basal lamina, or whether a lytic (enzymatic) effect has to prepare the perforation site. Cancer cell extensions frequently have an increased content of microfilaments and concomitantly of immunofluorescent labeling for actin (Gabbiani and others, 1976). This indicates the role of actin in producing localized cytoplasmic rigidity and contraction. Conjecturally, an extension projected through the basal lamina and firmly attached with its tip eventually pulls the whole cell across the membrane. This would be another transition from stationary to translocative cell motility during invasion.

All this information provided by static and dynamic techniques of morphology cells for an exploration of the mechanophysiology of cancer cell movement at the cellular and molecular level.

In general terms, such an exploration is part of a research area that has become very popular: the motility of non-muscle cells. No major difference between the machineries of cell motility of normal and neoplastic cells has been revealed so far. If we concentrate on locomotion, we can say that a cell moving on glass or another artificial substrate achieves propulsion by combining the generation of driving force with the construction of attachment sites. As concerns motive force generation, contraction of actomyosin, polymerization - depolymerization of tubulin and actin, and ill-explored membrane flow phenomena can all be involved, and a discussion of these molecular mechanims is beyond the scope of this paper. According to Wohlfarth-Bottermann (EORTC Workshop Proceedings, in press), the translation of motive force to locomotion shows a great divergence in different cell types. One factor responsible for this divergence may be the adhesive behavior. By means of reflection microscopy, great differences in the patterns of attachment sites of different cell types and of individual cells within the same population can be demonstrated (Haemmerli and Ploem, in preparation). At least in locomoting fibroblasts, sites of adhesion are decisive factors for the spatial organization of the contractile elements and thus for the conversion of motive force to propulsion (Heath & Dunn, 1978).

The principle of interaction between generation of driving force and adhesion can also be applied to the in vivo-situation. In most instances, however, attachment sites of cells moving in the organism do not have a two-dimensional, but a three-dimensional distribution, and this may be related to a phenomenon that is not or only incompletely recorded in most in vitro-studies on cell locomotion: the adaptation of the shape of the moving cell to its environment (EORTC Workshop Proceedings, in press). One gains the impression that many cancer cells are highly adaptable, but as the homologous normal cells do not display any translocative motility, a comparative statement is not possible. (To a certain extent, this is also true of leukemia, since in early developmental stages white blood cells are non-locomotive). Shape adaptation during locomotion is an active performance of the cell, and it is a reasonable assumption that it is accomplished by the same machinery that generates motive force . The two functional expressions can be expected to be closely correlated. In addition to active shape adaptation, passive modeling of a locomoting cell may play a limited role. It is more important during cell transport under the influence of a vis a tergo, e.g. in blood capillaries. Egress from capillaries can be passive, as in erythrocyte diapedesis, or active, as in emigration of white blood cells and cancer cells. Some data indicate that active adaptability and passive deformability are inversely proportional (Lichtman & Kearny, 1976).

A locomoting cell's possibilities to bypass obstacles by means of shape adaptations are limited. Changes of direction, short tracks, reduction of speed are constantly imposed on moving cells by the texture of their environment. This is clearly recognizable in the mesentery (Haemmerli & Sträuli, 1978). But even if all this is taken into account, the question remains in what tissue compartment the cells actually locomote. There are no empty spaces, and what appears as such in sections contains a network of molecular order, e.g. proteoglycans associated with collagen and constituting the extracellular matrix. It is undecided whether for propulsion in this material mere physical pressure generated by locomoting cells is sufficient, or whether small scale lysis of host constituents, not detectable by microscopic methods, is necessary. Apparently neutrophils, although well supplied with proteinases, do not need them for traversing tissue barriers (Armstrong, this symposium). For cancer cells, the joint operation of motility and lysis remains to be investigated. This is particularly important for an understanding of early phases of tumor growth in which single malignant cells may escape into host tissues. In later stages, conspicuous destruction zones often surround the tumor front, and this is not necessarily a situation favoring tumor cell locomotion, as in the lysed area support for attachment and nutrients may be lacking. In such instances, invasion is accomplished by the combination of proliferation and lytic action. In other situations, it is mainly based on the combined operation of proliferation and cell movement. Hydrolase activity may be a contributing factor, but for the time being this is more a theoretical postulate than an established fact.

REFERENCES

Sträuli, P., and L. Weiss (1977). Cell locomotion and tumor penetration. Europ. J. Cancer, 13, 1 - 12.

Le Douarin, N. (1976). Cell migration in early vertebrate development studied in interspecific chimeras. In Embryogenesis in mammals (Ciba Foundation Symposium No. 40), pp. 71-97.

Felix, H., G. Haemmerli, and P. Sträuli (1978). Dynamic Morphology of Leukemia Cells. Springer, Berlin Heidelberg New York.

Wood, S., Jr. (1958). Pathogenesis of metastasis formation observed in vivo in the rabbit ear chamber. Arch. Path., 66, 550 - 568.

Haemmerli, G., and P. Sträuli (1978). Motility of L 5222 leukemia cells within the mesentery. Virchows Arch. B Cell Path., 29, 167 - 177.

Haemmerli, G. and P. Sträuli. Have leukemia cells an inherent type of motility ? Leukemia Res., (in press).

Ambrose, E.J., and D.M. Easty (1973). Time lapse filming of cellular interactions in organ culture. II . Behavior of malignant cells. Differentiation, 1, 277 - 284.

Frithiof, L. (1972). Ultrastructural changes at the epithelial-stromal junction in human oral preinvasive and invasive carcinoma. In D. Tarin (Ed.), Tissue Interactions in Carcinogenesis. Academic Press, London New York, pp. 161 - 189.

Gabbiani, G., J. Csank-Brassert, J.C. Schneeberger, J. Kapanci, P. Trenchev, and E.J. Holborow, (1976). Contractile proteins in human cancer cells. Immunofluorescent and electron microscopic study. Am. J. Path., 83, 457 - 474.

Wohlfarth-Bottermann, K.E. Conformance versus divergence. In Proceedings of EORTC Workshop on Motility, Shape, and Fibrillar Organelles of Normal and Neoplastic Cells. Europ. J. Cancer , (in press).

Heath, J.P. , and G.A. Dunn (1978). Cell to substrate contacts of chick fibroblasts and their relation to the microfilament system. J. Cell Sci., 29 , 197 - 212.

EORTC Tumor Invasion Group: Proceedings of Workshop on Motility, Shape, and Fibrillar Organelles of Normal and Neoplastic Cells. Europ. J. Cancer , (in press).

Lichtman, M.A., and E.A. Kearney (1976). The filterability of normal and leukemic human leukocytes. Blood Cells, 2 , 491 - 506.

Index

The page numbers refer to the first page of the article in which the index term appears.